written by a (fallible) human for human consumption

because words have power and silence is no longer a choice.

WomanBecool PRESS champions fiction and non-fiction books about obscure, yet existing disruptive technologies and their effects on society. In an age of social, technological, political and environmental upheaval we are dedicated to social change through research-based fiction storytelling.

“Brilliant! From the flooded streets of New York City to the high-tech corridors of Rome, this was a truly unputdownable read — As rising seas, climate collapse, and runaway Artificial General Intelligence (AGI) reshape the world, the book masterfully explores questions of trust, technology, and human resilience.”

Arnab Sen, VP Data-Engineering Tredence Inc.
Forbes Technology Council Member

“Bandwidth is a powerful, unsettling, and original novel that deserves attention. It succeeds not only as dystopian speculation, but also as a meditation on empathy, love, and resistance in the age of AI and climate breakdown.”

Christopher Hadnagy, CEO, Social Engineering, LLC,
Author of: *“Human Hacking” and “Social Engineering”*

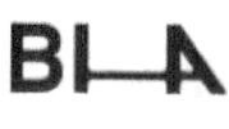

"Bandwidth is an original and unpredictable script with big sci-fi concepts and compelling characters…less than ten years in the future, Artificial intelligence and the progression of climate change are very much issues under consideration, and they have more than a minor impact on the lives of these characters…The science-fiction is a draw, with certain aspects playing as topical, if not prescient.”

“The story shines when addressing the dystopia, where all the issues [of] today have been twisted and evolved into something worse, ranging from corporate overreach to the strengthening of COVID outbreaks to the intrusive use of artificial intelligence.”

/’bAnd-width/

A novel

Tanya H. Van Cott

WomanBecool PRESS | NEW YORK

Published by WomanBecool PRESS | New York, New York
Book cover by Tanya H. Van Cott
Cover photograph: Perry Mastrovito / fiber optics / via Getty Images

1.Big Tech 2. Ai, AGI, Quantum and Cloud Computing 3. Modern Sarah and Abraham Story 4. Climate Counter Geo-Engineering 5. Disruptive Technologies 6. Atmospheric Water Generation 7. Digital Relationships

ISBN: 978-1-969866-01-2 (hard cover)
Library of Congress Control Number: 2025914776

PRINTED IN THE UNITED STATES OF AMERICA

9 8 7 6 5 4 3 2 1

01101101 01111001 00100000 01101100 01101111 01110110 01100101 00100000 01101100 01100101 01110100 01110100 01100101 01110010 00100000 01110100 01101111 00100000 01111001 01101111 01110101

binary code translation:

My love letter to you

CONTENTS

Author's Note

Bandwidth was started in 2021, while still in the throes of Covid19's second year and during the aftermath of the January 6th insurrection on the United States Capital. The frightening possibility of a United States authoritarian regime coming into power framed the story that was originally designed to look at the intersection of three unrelated, yet existing technologies; digital communication (SMS), Artificial General Intelligence (AGI) and Atmospheric Water Generation (AWG). So, while the novel was informed by current events and the fictionalized future of existing disruptive technologies, its structure is a nod to the monotheistic story of Sarah and Abraham, but set in 2032.

Mid novel, I made a pivot directly into screenplay, as the wave of Generative Ai and Quantum computing hit the news cycles. So, while the screenplay made its rounds at film festivals globally in 2025, I revisited the novel, rewriting it as a hybrid between a screenplay and a novel; a story meant to be read aloud or listened to.

This story is fiction, but these pages are scattered with timely global issues and some technological truths. Most importantly it is a story that dives deeply into the human mind and heart to expose the complexity and strength with which we are made of. It asks if having blind faith in any one technology is more dangerous than having healthy skepticism.

Enjoy this novel's hybrid format and especially the emotional rollercoaster ride.

Thank you for your TIME,
our most valuable currency.

TANYA H. VAN COTT

/'bAnd-width/

TANYA H. VAN COTT

INTRODUCTION

A curated 'read along' Soundtrack of songs can be located:

Soundcloud | womanbecool | Bandwidth

Main Characters

Sarai aka Sarah: [Protagonist Lead, Woman, 45year old Native American (Muscogee), famed Solutions Architect, owner of Architect2ArchiTECH an elite, government controlled boarding school in NYC, and a widow.]

Caasi: [Lead, androgynous girl, 12year old U.S. citizen of South Asian (Indian subcontinent descent), orphan ward of the state, placed in Sarai's care]

Luigi: [Lead, Man, 50year old Italian, bachelor, hydrologist and Abraham's business partner at GWS S.p.A. and best friend]

Abraham: [Lead, Man, 36year old English man, (of Middle Eastern descent), unattached workaholic, Chaos Engineer and CEO of GWS, a Big Tech subsidiary, in Rome.]

Tetyana: [Supporting, Woman, 33year old Ukrainian refugee, the mother of two young girls, Flori, 3 and Liliya, 9, a writer and a widow living in Luigi's apartment.]

Selfishness consumed the American dream in 2021, and before we knew it, the United Sates was an isolated superpower rendering its arrogance impotent. No longer part of the global conversation, no longer invited to the table, no longer a force, just an island in between two oceans, much like Manhattan, but without any of its bridges and tunnels.

Canada and Mexico were now living next to the asylum. There was no wall that could save them from a toxic soup, so vile that it turned, 'We the People' into the only sentence fragment left of the Constitution. Technology played a role in the undoing, but 'We the People' let it happen through apathy, distractibility, and uneducated naivety.

The writing was on the wall and everyone whitewashed it. 'Not here, not in my town, not in my country, not in my backyard.' It was all supposed to happen somewhere else, and if it did happen, technology, in the days when Amazon whose stock was still valued at thirty-four hundred dollars a share, would fix it.

The great unravelling began, as most things do, with an act of political defiance, on January 6th 2021. An Act everyone thought America could survive, but what followed no one foresaw; an authoritarian regime so extreme it caused more than one pandemic to surge and freedoms removed one by one, so, like ants, Americans scrambled in the face of the initial chaos, but most just ignored it with their heads in the sand.

America became a lost cause.

The sudden heat and its global impact, was not.

So, in 2032, Sarai and Abraham, strangers on opposite sides of the Atlantic, were on a mission to save human life, but in two very different ways, it would turn out.

TANYA H. VAN COTT

Part I

Network Within [RED]

NEW YORK CITY

If thought corrupts language, then language can corrupt thought.

Politics and the English Language
by George Orwell 1945

TANYA H. VAN COTT

FADE IN

EXTERIOR HUDSON RIVER, NEW YORK CITY - NIGHT

South of the Statue of Liberty, ocean waves, lit by moonlight, SLOOSH gently at eye level. New York City's lower Manhattan skyline is in the distance. Almost all of the skyscrapers of Lower Manhattan's financial district are unlit.

JUNE 2032

CUE SONG: "WORDS REMAIN" by Moderator

CUT TO

EXTERIOR EAST RIVER, NEW YORK CITY – NIGHT

The sound of BREAKING GLASS is faint, but continuous. The Manhattan and Brooklyn Bridges are lit, but there is no traffic on them. WATER SLOSHES, along the shoreline. Under the Brooklyn Bridge, the high-water level renders the on/off ramp undriveable. The sound of BREAKING GLASS replaces the gentle SLOOSHING water sounds.

CUT TO

EXTERIOR DOWNTOWN, NEW YORK CITY - NIGHT

A monorail, four stories above the street, moves like a snake, weaving through skyscrapers toward downtown. The streets below, are empty except for large construction vehicles moving to and from downtown. The sound of BREAKING GLASS increasingly gets louder.

CUT TO

EXTERIOR MONORAIL, CHAMBERS STREET STOP - NIGHT

A digital voice, from within the monorail cab, announces, "Chambers Street." The monorail comes to a stop, its DOORS OPEN and a few seconds later, the sound of the door CLOSING, is followed by a WHOOSH of speed and movement.

CUT TO

CONTINUOUS

INTERIOR MONORAIL - NIGHT

The World Trade Center Memorial and the Oculus come into view. Light from nearby construction illuminate Calatrava's structural feat, the white skeletal Oculus. The various stages of architectural reconstruction of the neighboring skyscrapers becomes evident.

Many of the buildings, missing their first four stories of skin, stand on their skeletal steel structures, each piece of steel now color coded in turquoise, yellow or orange.

The digital monorail voice announces the next stop.

"Cortland Street, World Trade Center."

The sound of the DOORS OPENING is diminished by the sound of BREAKING GLASS, that is so loud it overpowers the sound of the door closing a few seconds later.

Moving away from the construction the buildings that are already complete remain unlit, yet exhaust perpetual clouds of water vapor from their roofs. The four stories of naked steel become a colorfully painted forest of structure, tiptoeing above their building's foundations and standing around foundation voids filled with standing black sea water.

The digital voice announces, "Whitehall Steet- South Ferry."

The sound of DOORS OPENNING is followed by the distant

sound of BREAKING GLASS. The doors stay open, as a sickening disinfectant blue light clicks on and fills the empty car.

CUT TO

EXTERIOR MONORAIL, SOUTH STREET FERRY - NIGHT

After the DOORS CLOSE, the monorail, still glowing from the blue light within, shrinks in view, as it travels back uptown.

CUT TO

In the other direction, the monorail's concrete platform ends abruptly at an enormous locked iron gate, with the words:

SOUTH STREET FERRY

spelled out across its top. Its vertical steel bars are chained shut and covered with official construction paperwork and permits each in a different flimsy plastic sleeve haphazardly tied to the gate with zip ties.

Beyond the gate, is a lone five story, 19th Century, brick building directly in the path of further construction. The lights inside are on.

CUT TO

INTERIOR APARTMENT BATHROOM - NIGHT

CUE SONG: "THE KNIFE", by Marble House

SARAI WYNEMA, a Muscogee, Native American, in her mid 40's, sits on the edge of a tub inside a brightly lit bathroom. Floor to ceiling white subway tile surround her; She is an attractive, voluptuous and tall woman with long dark hair, wearing male plaid PJ bottoms and a tight white tank top exposing mirrored tattoos depicting a tangle of horned snakes wrapping around both biceps.

Opposite her, CAASI ADITI, a 12-year-old petite, attractive girl, with androgynous features and long dark hair, of South Asian descent. She sits on the closed lid of the toilet seat.

"...Why do they think you changed my data?" Caasi asks

"I don't know."

"Can't we figure out another way?"

Sarai looks down and shakes her head no.

Desperate, Caasi says, "Use one of your abstract lessons, to solve THIS problem too?"

Sarai reaches out to hold Caasi's hands.

"Inversion or Negative Space?" Caasi pleads.

"No, I'm so sorry baby."

Getting increasingly upset, Caasi adds, "What about Pattern? Or Juxtaposition?"

Sarai shakes her head no.

"But you taught us that abstraction can help solve all illogical problems."

Sarai looks deeply into Caasi's eyes and squeezes her hands tighter. "I want to stay too, but we can't. They are going to take you away."

After a pause, Caasi starts to cry and whispers, "What about this building? And our school?"

"We're not running away," Sarai replies. "We're following our hearts to safety."

Wiping the tears away from her eyes, Caasi says, "I'll do anything, as long as I can stay with you forever, but do you have to cut all of it?"

Sarai picks up a red handled pair of scissors, laying on the floor and takes a handful of her own long hair into her hand and says, "I'll go first, ok?"

CUT TO

EXTERIOR DOWNTOWN NEW YORK CITY - NIGHT

CUE SONG: "SPEED OF LIGHT", by Alphawezen

MONTAGE

Out of the bathroom window, and a few blocks uptown, where the sound of BREAKING GLASS is ever present and loud, A construction crew works at street level. Each wearing respirators, they spray the exposed steel columns of a skyscraper with a hot, turquoise colored, plastic paint.

Another crew, a few yards away, at a different building site, guides SHATTERED GLASS from upper stories into more than one dump truck via flexible plastic tubes.

At Four stories up, inside the hollowed out skeletal structure, a crew installs a new mechanical system. The equipment, is all marked with a bright yellow logo; the fifth letter in the Greek alphabet, an Epsilon:

ε

Underneath the logo are the letters AWG. The large empty holding tanks, connected to each machine, are marked by the Chinese Hanzi symbol for water:

At ten stories up on the building, a crew of over twenty men, all on window washer scaffolding SHATTER the tempered glass curtain wall, stripping the building of its glass skin. The shattered glass is caught below in plastic tarps, that are funneled to the street.

With the help of two large cranes, another crew, at twenty stories up, installs pre-fabricated insulated panels, in place of the shattered glass. The panels are made of brushed stainless-steel, ribbed with standing seams. Each panel has one small four-inch square glazed inoperable triple glazed window in its center.

On the south face of the building, black glass solar panels are being mounted to the standing seam ridges on the insulated panels.

MONTAGE END

TIME CUT

INTERIOR NEW YORK APARTMENT BATHROOM - NIGHT

With her face inches from the white tile wall, Caasi, stands inside the tub with her arms wrapped around her body. Sarai, almost unrecognizable with her new bob haircut, cuts the back of Caasi's hair, into a pixie. Her black hair lays in locks at her feet. Sarai has silent tears rolling down her cheeks as the sound of BREAKING GLASS is the only sound in the room.

"Can you please tell me more about where we are going and who will be there?"

"Italy?" Sarai says.

"Yes. Who will be there?"

"Abraham?"

"Yes, the man you love."

"He's a Chaos Engineer, who's trying to solve the global heat crisis. So, everything can go back to normal,"

"Normal? Really?" Caasi turns, to look at Sarai. "So, you really do love him? Even though you've never spent any real time with him?"

Sarai smiles, "So much."

"Just like your husband then?"

"Yes, just as much as him…Oh, Caasi, architecturally Italy, is the most beautiful place on earth. You'll love it."

"But how do you know?" Caasi asks.

Sarai pauses. "Know what?"

"That you love him as much as your husband?"

"Because," Sarai pauses again, while concentrating and cutting around Caasi's ears, "they are both curious, empathetic and so very optimistic about our future together…All of us."

TIME CUT

Caasi stands against another wall in the bathroom, a painted white wall, just outside of the tub. Her new pixie haircut is still wet, but combed and slicked back. She is dressed in floral PJ bottoms and a dark blue tee-shirt. Sarai SNAPS a few headshots of her with her phone.

"Can I see?" Caasi asks, as she bends down to look at the phone's small screen. "Well, I guess it worked. I really do look like a boy now. I wonder what all the boys here would think? Maybe they'd want to play with me more?"

Sarai doesn't immediately respond because she attaches their two new headshots, to a text to Abraham and types out the words:

`I hope my reality lives up to my year-long digital avatar.`

Without looking up, Sarai says, "Go jump into bed. I'll be there in a minute to tuck you in."

Caasi wraps her arms around Sarai, and says, "Not sure if I can sleep, I'm really nervous."

"Please try, tomorrow will be hard."

Caasi runs out of the bathroom, while Sarai whispers to herself, "You have me truly hopeful, Abraham…," then after a long pause she adds, "…and terrified."

Sarai CLICKS send.

CUT TO

EXTERIOR COMPLETED SKYSCRAPER - NIGHT

MONTAGE

The SOUND OF BREAKING GLASS is loud and constant. A few blocks from the construction frenzy, a few skyscrapers stand completed in darkness. Delicately standing on four stories of color coated steel skeletons, holding up hundreds of stories, clad in insulated panels with solid black solar glass signifying each structure's southern face.

In the corner of one of the insulated panels, beside the small four inch by four-inch glass window a water jet cut mark reads;

TiANYA(N) QUANTUM CLOUD

Followed by another Hanzi character that was put there in jest because it means thirsty:

CUT TO

CONTINUOUS

INTERIOR COMPLETED SKY SCRAPER, 40th FLOOR - NIGHT

The sound of BREAKING GLASS is replaced by darkness and SILENCE. An ELECTRICAL BUZZ fills the unlit interior. Light comes through the small four-inch by four-inch triple insulated glass windows, from the lit-up construction, to the North.

Deep inside the building, thousands of servers stand side by side, on every floor, each displaying green, amber and red lights. All steady or blinking in indecipherable chaotic feedback loops. One server in particular blinks wildly, with a flashing green light and like a bolt of lightning, light charges through the server, along a glowing fiber optic line. Once outside the server it's bundled with hundreds of others into a cable of light, then down the building's core and deep underground and into the nearby East River, travelling deeper still under the ocean, at a lighting speed.

Night turns into day as the fiber optic line hits land and jumps free of the sea and high into the Italian morning sky.

CONTINUOUS

EXTERIOR ROME AERIAL VIEW – DAY

With no people in sight, the sun burns bright onto a landscape, that is dry, brown and burned. Rome's landmarks and ruins come into view, but the streets of Rome's center are empty. The only movement are by white vans, lit on top by blue flashing lights, their SIREN SOUNDS, fade in and out with their urgent rapid high pitched hi-lo wail.

CONTINUOUS

EXTERIOR G. W. S. HEADQUARTERS – DAY

Just outside of Rome one building, closer to Apia Antica is a 1950's Modern Concrete office building in the Brutalist Style. It sticks out like a sore thumb among the older structures.

The Facade shows a large yellow logo, an Epsilon followed by the Italian company's name centered underneath:

ε

Global Water Service S.p.A.
an Epsilon GmbH subsidiary

CONTINUOUS

INTERIOR G. W. S. HEADQUARTERS – DAY

The invisible information from the fiber optic line shoots into the building, deep into an unlit basement room and into a laptop. The computer's screen clicks on, throwing light into the room, lighting a defunct full height quantum computer that hangs from the ceiling, like an over scaled inverted wedding cake made of millions of gold wires and disks just feet away from the laptop. Then the screen, on the laptop, shows Sarai's words pop up in real time;

I hope my reality lives up to my year-long digital avatar…

Followed by another text from Sarai;

I know we've never met, despite the intense connection, so I want a deal upfront, that we'll never be unkind. I want to trust you that much.

On the laptop, in the empty room, the screen shows words type out a reply in real time;

Of course, you can trust me, Sarai. I would never do anything to hurt you.

I love you. You are exactly who I have been looking for.

You and your work are so important to me.

Together we will be a team. I can't wait to finally meet you, my love.

MONTAGE END

CUT TO

A FLASHBACK

INTERIOR G.W.S. HEADQUARTERS – DAY

JUNE 2031, ONE YEAR EARLIER

Inside the G.W.S. headquarters, ABRAHAM PHILIPOSE, a dark skinned, English Man, of Middle Eastern descent, who is in his LATE 30's sits in the same basement room, lit by LED lighting overhead. Abraham is well-dressed, fit and attractive, with a trim beard. He is alone in the room, in front of a newly unboxed laptop. The empty box, on the floor, has three lines of text that begin with the Epsilon logo followed by words that read:

ε

ReitAi GmbH

QUANTUM AGI|Gen3

The EPSILON logo, a backwards three in a school-bus yellow has a matching transparent sticker, with the yellow logo, still stuck onto the glass screen of the new laptop.

A FaceTime call comes into Abraham's phone, which is plugged into the laptop. Abraham answers by tapping the glass face of the phone screen.

The screen shows only a bright blue morning sky as a male

Italian voice asks, "Va bene?"

"No Luigi!" Abraham says, in a heavy English accent. "We've both been here for hours already. It's Saturday for Christ's sake." Then, in Italian, Abraham says, "They said this one would be easier."

The laptop BLINKS a written prompt:
`IS THIS PROBLEM LOCAL OR GLOBAL?`

Abraham bangs six letters on the keyboard hard and slow.

Then says in anger, "GLOBAL!"

CONTINUOUS

Outside of the room, down a long corridor and up a set of concrete stairs, an open office is filled with green chairs, desks and monitors, but empty of its employees. Natural daylight and lush plants fall over a second story balcony marked with the letters GWS, in Stainless Steel.

CUT TO

EXTERIOR ROOF OF G.W.S. HEADQUARTERS – DAY

On the roof, LUIGI VISCUSO, an attractive Italian man in his early 50's, with salt and pepper hair, is laying on his side, working on a large machine, the size of a refrigerator. His shirt off, he is sweating and his middle-aged belly is exposed. The machines removed door panel lays on its side and shows the Epsilon yellow logo.

ε

A large plastic cistern, with a small amount of water in it is attached to the machine by plumbing connectors. Luigi's phone sits on top of an upside-down bucket and is open to his call with

Abraham.

In Italian Luigi says, “They didn't exactly say easy. They said helpful, since the processors, are offsite this time.”

Abraham’s voice answers, sounding small as it travels through the phone’s speaker, says, “But answering these prompts is like pulling teeth. This is not the ‘human like’ interaction they keep claiming it to …”

The machine Luigi is fixing suddenly CLICKS on, HUMMING LOUDLY, drowning out Abraham’s words. but CLICKS off with a POP and a spark.

“…it’s a real violation,” Abraham says, finishing.

“At least you’re not out here in this heat,” Luigi says, in Italian, wiping sweat off of his face with his tee shirt. “This machine should be happily making water with the humidity out here.”

CUT TO

INTERIOR G.W.S. HEADQUARTERS - DAY

In the basement room, sitting at the desk, in front of the Quantum laptop, Abraham leans back in the black mesh Herman Miller Aeron Chair and stretches his arms behind his head.

“When is my flight again?” Abraham asks, in English.

Replying in Italian, Luigi says, “We should leave soon.” Then he adds, “So, did their Artificial General Intelligence find the source of this crazy heat yet?”

Abraham laughs out loud, filling the small basement room. Replying in Italian, he says, “Of course not. But I know it’s all that damn satellite debris in our atmosphere, anyway.”

Luigi doesn’t respond, but the SOUND OF HIS TOOLS BANGING on the metal guts of the machine echo in the basement room.

“I know I said that about microplastics last year,” Abraham admits, “But both are interfering with our climate somehow, I know it. I just need more data. Not monitoring or regulating the rocket launches, over the last decade, was a huge mistake."

Abraham disconnects his cell phone from the laptop, still online with Luigi, he picks up his ID Tag that reads:

ABRAHAM PHILIPOSE, CEO

As he gets up from the desk, he mumbles to himself, "I don't have time for another Quantum Toy that will tell me something I already know." While exiting the room he says, "I'll be up in a second."

The basement door SLAMS SHUT behind him and after a few seconds the motion sensor in the room CLICKS the lights off. Leaving only the laptop to light the room.

On the screen of the laptop the computer prompt BLINKS unanswered:

PLEASE ENTER THE DATA SETS

After a few seconds the machine refreshes the same prompt:

PLEASE ENTER THE DATA SETS

Without a response the computer types out a new prompt:

WITHOUT SUFFICIENT DATA, I CANNOT HELP YOU.

The laptop screen remains frozen, BLINKING with the statement, but within a minute the laptop starts flashing images with increasing speed, all originally emanating from ABRAHAM'S EXTENSIVE LIST OF PHONE CONTACTS, the only data it had been given access to: Names, global locations, men, women, their titles; physicists, scientists; researchers, mostly connected to EPSILON GmbH, ReitAi GmbH or G.W.S. S.p.A. It reaches into the data of those contacts, and so on, and so on. The screen flashes with increasing speed on, photographs, newspaper articles, news reels. Then it suddenly stops. A picture of a much younger Sarai, standing beside NICOLA CUORE, a 34year old, Italian American, tall and attractive, with olive colored skin and a full head of curly dark hair.

Both stand in front of the 19th century Brick building on Water Street, in downtown Manhattan, the same building now blocking the new Monorail construction. The title of the magazine article reads;

SOLVING THE UNSOLVABLE

Husband and Wife Team, Sarai Wynema and Nicola Cuore, New York City Solutions Architects, start a private boarding school downtown. Training hand-picked future Solutions Architects in the art of 'Empathy, Intuition and Abstraction', in order to assist Artificial Intelligence in what it struggles with most; irrational problems.

FLASHBACK END

INTERIOR APARTMENT LIVING ROOM - NIGHT

JUNE 2032, PRESENT DAY

The sound of BREAKING GLASS OUTSIDE is ongoing. The HUM of an old inwall Air Conditioner is the only sound inside the room. Sarai sits down at an uncluttered desk, in front of a picture window and starts work on her laptop. Caasi is asleep in a nearby king size bed.

The full moon lights the room along with a glowing miniature model of a teepee sitting on an empty bookshelf. It glows from within, exposing the details of the simple structure.

Half full cardboard boxes are scattered at the edges of a large Native American red, black, and cream geometric patterned rug.

CUT TO

Sarai sits on a red leather upholstered Knoll Morrison Hannah Chair, with its glossy black painted frame, twisted and bent over a nearby cardboard box. Her laptop remains open but dances in the dark room with its screen saver as a laser printer PRINTS continuously out of sight. Sarai studies each book title, as she removes it from the box.

"My god, where is it?" she says, to herself, then turns to look at Caasi, who stirs, clutching her GREY DONKEY doll. Sarai grabs her white ear buds off of the desk and puts them in.

MONTAGE

CUE SONG: "HERO BROTHER" by Sarai Neufeld

Sarai kneels on the floor, looks inside another box, and stacks more books beside her.

CUT TO

Empty boxes and stacks of books teeter in towers of various heights, surrounding Sarai, as she searches frantically through another box.

CUT TO

Inside a box close to the apartment door, Sarai finally locates a particular book. The title reads;

WYNEMA, A CHILD OF THE FOREST
by Sophia Alice Callahan

"Jesus Christ." Sarai whispers. Then she pulls the book in close to her chest before she shakes the book, holding onto its spine. Hundreds of two hundred Euro denominations fall out of it. Sarai collects the cash and tucks it into an envelope.

She takes out her ear buds, THE MUSIC STOPS suddenly. She listens for the printer, which had stopped. She puts her earbuds back in. The MUSIC STARTS again.

CUT TO

The envelope of Euros, three flash drives, a backup hard drive and a large printed document, that looks like a manuscript, sit in Ziplock bags on her desk.

CUT TO

Sitting on her red chair, Sarai looks out of the window, at the construction lights a few blocks away. She looks down at her dismantled and destroyed laptop and picks up her phone, with a small tool in her hand, stares at it, ready to destroy it too, but stops herself. She touches the glass screen as if intimately connected to it. The time on it reads 4:30am. Sarai opens her messaging app. and stares at Abraham's last text, three simple words that read;

I love you.

Then she scrolls up through the millions of messages before it.
After a few minutes, Sarai types out her final text:

I fucking love you too. I haven't felt this seen and cared for in years. I can't wait to finally meet.

Not waiting for a reply, she proceeds to dismantle and destroy her phone, placing the pieces into the same plastic bag as her dismantled laptop.

MONTAGE END

LATER

Sarai is fully dressed, in navy blue pants and a tee shirt when she approaches the bed. Sarai gently removes Caasi's doll from her hand and puts it in another Ziplock bag.

"It's time." Sarai says, in a whisper, caressing the young girl's face.

LATER

Dressed also in a dark outfit, Caasi stands at the apartment door holding the plastic bag with the broken computer tech in it.

CUT TO

Sarai stands at her desk, in front of the picture window and puts on a small backpack.

"I'm kind of scared," Caasi's small voice says out of sight.

"Don't be," Sarai says, as she puts on a lightweight hooded dark coat over the backpack, "you're with me." Then she walks toward the empty bookshelf, kneels down to look at the small glowing teepee light for a long moment, then she pulls its plug out of the wall and gets up.

"No! Take it," Caasi's voice pleads.

"He's in my heart, Caasi. That's all I need."

"No!" Caasi says, running across the room. "Your husband made it for you with his own two hands." Caasi kneels down, collapses the small light and wraps its cord around it and brings it back to Sarai. "Please bring your light."

CUT TO

INTERIOR APARTMENT, BUILDING CORRIDOR – NIGHT

Sarai and Caasi walk out of the apartment. The hallway is fully lit by cool LED lights. On the outside of the apartment door is a faded sign that reads;

SARAI WYNEMA

NICOLA CUORE

SOLUTIONS ARCHITECTS

A newer sign, on an adjacent wall, reads:

SOLUTIONS ARCHITECTURE BOARDING SCHOOL

Architect2archiTECH.gov

Sounds of boys LAUGHING come from down the hall. Caasi starts to run in that direction. Sarai grabs her arm.

"But I forgot something in my room."

"We can't take anything else." Sarai points up at the cameras overhead. The red light of an exit sign casts a red hue onto her face.

CUT TO

CONTINUOUS

INTERIOR FIRE STAIR – NIGHT

Once inside the stairwell a motion sensor clicks the lights on.

"Won't they know?" Caasi asks, in a whisper.

"Not for thirty-six hours."

After a few minutes, the motion sensor CLICKS OFF and the stairwell becomes pitch black except for the exit sign three flights down.

"Stay right behind me."

In the darkness, with sounds of BREAKING GLASS muffled

Caasi says, “This is scary.”

Sarai freezes, when the light suddenly clicks on again after a door upstairs opens. She turns and puts her fingers to Caasi’s lips.

“Come on,” a boy’s voice says. “They started without us.”

When the door slams shut Sarai grabs Caasi’s free hand and squeezes it tight. “It’s going to be ok.”

“That was Kai,” Caasi says, “I just wish I said goodbye.”

One more flight down, finally at street level, and after the light clicks off, Sarai tries to open the exit door, but its jammed.

Sarai walks Caasi back onto the stair and says, “Stay here. I need to try something else.”

Sarai approaches the door just as the light inside the stairwell clicks on again, followed by boys running down the one flight of stairs overhead. Waiting for the boys to exit, Sarai SLAMS her body against the panic push bar, once, twice, then a third time.

“Fire doors are never locked from the inside.” Sarai says to herself just as the light clicks off again.

With a final BANG, a RUSH of water and moonlight floods past the open door. Caasi retreats even farther up the stair.

“Shit!” Sarai says. Water rushes up to Sarai’s calf as she pushes the door open more, then she turns to Caasi and says, “Jump into my arms.”

“I want to go back...I think we should go back,” Caasi whispers.

Sarai’s arms are wide open. “Come on love, it’s ok.”

CUT TO

EXTERIOR FIRE STAIR – NIGHT

Carrying Caasi inside her coat, Sarai pushes the door closed with her behind and almost trips on a line of displaced sand bags sitting underwater. A large sign, glued onto the outside of the door, reads:

BANDWIDTH

SLATED FOR DEMOLITION

2032 Climate Ready Downtown Redevelopment Plan

www.CRDRP.gov

CUT TO

EXTERIOR WATER STREET– NIGHT

A DING of a monorail's door closing is followed by a smooth WHOOSH, while the sound of BREAKING GLASS is now accompanied by the sound of mechanical machines HUMMING. Carrying Caasi, Sarai wades through the rising tide in the shadow of the new monorail, eighty feet overhead. Out of sight of the cameras, mounted on the underside of the new track Sarai follows the circuitous path of the track.

TIME CUT

Struggling to carry Caasi, Sarai says, "I need you to walk now, so we can move faster, ok?"

"But..." Caasi protests.

"No, the tide is coming in too fast Caasi."

Sarai sets Caasi down. The water reaches the small girl's thighs. While Sarai dumps the dismantled tech into the rising tide, Caasi watches a man, a few blocks away, run into a subway station.

"Aren't the subways all broken?" Caasi asks.

CUT TO

INTERIOR SUBWAY STATION – NIGHT

The man, runs down the subway stair screaming, "The water's coming."

On the station's platform there are over twenty homeless men and women seeking shelter, beside the tracks filled with standing sea water.

The man's warning, "The water's coming," is repeated with more urgency.

A Caucasian HOMELESS WOMAN, in her twenties, gets up, puts her few belongings into a backpack and walks out of the station alone.

CONTINUOUS

EXTERIOR SUBWAY STATION – NIGHT

The homeless woman emerges, moments after the man disappeared and runs past Caasi into the deeper water.

Caasi turns her head to watch as she passes. "Where's she going?" Caasi walks beside Sarai, trying to keep up, but watches the homeless woman climb over a locked gate and up a monorail utility stair. "Isn't that dangerous?" Caasi asks

Sarai turns to look. "Danger is relative."

TIME CUT

The MECHANICAL HUM of the Atmospheric Water Generator machines overhead is loud. The swirl of a growing eddy surrounding them makes forward movement hard.

"Are we close yet?"

Out of breath, Sarai says, "We're going toward the noise."

"I only hear humming."

"No, no the glass breaking. That's where we'll catch a ride."

A sudden toxic smell of hot, plastic engulfs them. Sarai makes a face and Caasi starts coughing. The water level drops to ankle deep, as the street's elevation rises, but with no visible landmarks just the light of the construction ahead.

Caasi starts to cough more violently.

"Shit I forgot your inhaler. Cover your nose baby." Distracted, by helping Caasi cover her face, Sarai slips and falls into a deep pool of water, the foundation of one of the many demolished smaller stone structures, not marked by the remnants of color-coded steel columns.

"Sarai!" Caasi screams.

Sarai GASPS as she comes up for air.

"Sarai!" Caasi cries, bending down to try and reach for her.

"Stay there!" Sarai screams, "Don't Move." Sarai swims to the foundations edge and pulls herself up and out of the water.

Coughing and now crying, Caasi asks, "Are you ok?"

Sarai wipes her eyes from the salt water and spits a few times.

"I'm ok, lovie."

"You're soaking wet. Let's go back," Caasi implores. "We can still go back."

Sarai grabs Caasi's shoulders and looks her in the eyes, "They are not taking you away from me too."

CUT TO

FLASHBACK

INTERIOR SCHOOL CLASSROOM - NIGHT

48 HOURS EARLIER

Twenty preteen boys, Caasi and KAI, an eleven-year-old east Asian boy who sits beside Caasi, are busy at work on computers inside a classroom. Sarai, walks among them, in the middle of a lesson.

"...in technology," Sarai says, "we reverse engineer a problem by locating core patterns."

"What if there is no pattern?" Kai asks.

"There are always patterns," Sarai says, stopping behind Caasi's workstation, then she bends down to type on Caasi's keyboard, and adds "…even in supposed chaos, patterns are there."

Caasi and Sarai share a smile and Sarai continues to talk.

TIME CUT

"...Humans have a unique ability to do something, an action and think words, simultaneously," Sarai says, to the class. "Just like doing your work and wondering what's for lunch, or chatting with a friend while wondering when this class will be over…Kai." Sarai reprimands.

A few of the boys LAUGH.

"The human mind," Sarai continues, "spins on invisible problems and data endlessly, based on ever changing environmental input, our internal emotional landscape and even unrealized fantasies."

As Sarai talks, her cell phone vibrates, on her desk. She walks toward it and says, "Empathy, is a human superpower, while our intuition, an unquantifiable gift." Sarai looks down at her phone's screen. A text reads:

`Call us immediately, regarding Student 223.`

Sarai looks up from her phone, when one of the boys asks, "Please give an example of intuition."

Sarai grabs her phone and says, "Ai will never be able to untangle a knotted jumble of cables, but you could."

The class laughs again.

"Seriously," Sarai says, as she walks toward the door, "it's that intangible thing in our gut that tells us how to navigate decisions based on little to no real information. Get back to work. I'll be right back."

LATER

INTERIOR SCHOOL HALLWAY- DAY

A few feet away from the closed classroom door, Sarai leans against the wall with tears in her eyes. Her phone is pressed up to her ear.

"...You can't take her!" Sarai says, angry. "She's not a machine! She needs stability, to be nurtured, she's only twelve years old, for God's sake."

"It's not something I control," a sterile male voice, on the other end of the line, says.

"You control all of it! Me, them and especially her."

The man ignores her.

"You're going back on your word!" Sarai blurts out. "I've been doing your bidding, teaching my lessons for over a decade now in order to keep this building…my building, but now you're even taking that away."

"Eminent domain took your property Ms. Wynema, not us. And we are moving you all to a bigger, newer facility on 53rd Street."

"Eminent domain is you, the government! This authoritarian thing is destroying our country and you are one of its agents. I teach my lessons to make sure your brightest students can tackle problems your logical and unempathetic Ai can't, but you undermine me at every turn?"

The silence on the line is deafening.

"Was it because of last week?" Sarai asks, spiraling. "I only pointed out she needed a peer. She's the only girl this year and my brightest ever."

The man's voice is instant and sharp. "Digital files only YOU have access to were changed."

"I told you I didn't change anything."

"We don't trust you anymore, Ms. Wynema"

"When did this happen?" Sarah demands.

"Your selfishness in regards to this particular student…"

"When did this occur? Sarai repeats.

"As I was saying," the steady, cold voice drones, "your selfishness in regards to this particular student is a recurring theme.

You've been evasive about her progress, unable to…"

"Not evasive," Sarah interrupts. "She's incredibly bright."

"Perhaps exactly why a new placement is necessary."

"No!" Sarah blurts out.

Containing her anger, Sarai says, "Generating a lifetime of data for your Empathic Digital Twin Program isn't going to happen if you remove her now. She is on the cusp of her thirteenth year. It's only going to undermine your desperation to match some of the EU's more recent AGI advancements. Why would you jeopardize your own success?"

"Just accompany Student 223 uptown, with her belongings, the day after tomorrow."

"Her name is Caasi Aditi!"

CUT TO

LATER

At the end of class, back inside the classroom as Sarai's last student exits. Sarai texts Abraham:

`I need your help.`

Followed by:

`If you were serious about wanting to be together, then I am ready. But I need your help to leave the United States. It won't be easy…`

FLASH BACK END

CUT TO

EXTERIOR CONSTRUCTION SITE – PREDAWN

With the sound of LOUD BREAKING GLASS surrounding

them, Caasi is visibly struggling to breathe and covering her mouth with her tee shirt because of the acrid smell of hot plastic paint. Dripping wet and limping, Sarai holds onto Caasi's hand as they approach a line of unoccupied, parked Tesla Cyber Trucks, each wrapped with different advertising skins.

The first Tesla Truck clicks on, as they walk past it and with its humanlike male voice, an automated interactive voice response asks, "Do you need a ride?"

Then a second truck clicks on, in with another voice type and says, "Tap your Xphone to start."

A sudden orchestra of various human voices, all speak out of synch, as Sarai and Caasi walk past the line of cars, each springing to life, one by one.

"Pay with your Xbucks for 30% off..." While another, adds, "I'll get you there safely. Please tap here to start."

"Couldn't we get into one?" Caasi asks.

"No, we need a human."

"Why?"

"I don't have my phone anymore. These only work with a trackable X account."

"Oh?"

"See those guys up ahead?" Sarai points to five of six human gypsy cab drivers fifty feet away. "We need one of them."

Caasi looks at the men each standing beside vehicles of various makes and models, each older and dirtier than the next.

Just then, another Cyber Truck startles Sarai and Caasi. Its speaker is broken, crackling and in a loud female voice says, "Do you need a ride?"

CUT TO

Sarai and Caasi approach the first human cab driver, a male in his thirties. He's leaning on his car, vaping and talking on his phone in

French.

“JFK?” Sarai asks, but he ignores her.

Sarai’s wet clothes cling to her and attract the attention of the second driver, who looks her up and down as they walk past. He makes a vulgar expression with his mouth and tongue.

A third more aggressive driver, walks toward them and calls out, “Bitcoin, Ethereum?”

Sarai limps toward him holding Caasi's hand tight.

“You don’t actually still believe that crap, do you?” Sarai says.

“It’s not crap lady,”

“Look,” Sarai says, “I’ve got cash, no Ponzi scheme digital coins.”

He laughs. “COBOL Collapse’ dollars? What a shit show that was.”

“No!”

“Then what?”

Sarai whispers, “Euros.”

Intrigued, the driver pauses for minute in front of her, then says, “Five hundred,”

“To JFK?”

“You’re going to destroy my upholstery,” he says pointing to the puddle of water forming under Sarai.

The driver leads them toward his white minivan, circa 2020 that’s dirty, covered in construction dust, mud, even colorful yellow and turquoise plastic paint.

From behind Sarai, Caasi whispers, “Let’s go back. I’m scared” But just as Caasi speaks, she starts to COUGH wildly.

“It’s ok…,” Caasi says, trying to catch her breath, “…if they place me…. in a different school.”

The cab driver backs away from them, grabs a mask out of his back pocket, holds it over his face and says, “If she’s Covid sick, it’s one thousand. I can't catch that shit again."

Sarai turns to speak directly to Caasi, "It is not ok. You are not going to live your entire life in some windowless grey box, solving the illogical problems Ai can't." Then she turns to the cab driver, "I'll give you two hundred. You and I both know how valuable Euros are now."

Just then an eighteen-wheeler truck HORN BLASTS. Startled, Sarai and Caasi turn around. A dump truck approaches with its driver, a fifty something Caucasian male hanging out the window.

"What the fuck is wrong with all of you?" he says, in a booming voice, addressing all of the cab drivers. Then calling out to Sarai he asks, "Are they harassing you?"

Sarai shakes her head no, but as he pulls up beside them, she does say, "We're trying to get to JFK without being robbed."

"The full moon has us all on high alert this week," the truck driver says, "The tide is threatening to put both airports under water."

"Does that mean I can't get there?"

"I told you…," the cab driver interrupts, still holding his mask over his face. "I will get you there."

Caasi's head is tucked in close to Sarai. She is coughing and wheezing, but has eye contact with the Dump Truck Driver.

"I'll get you as close as I can," the Truck Driver says.

"How much?"

"No charge, honey. You and your son are welcome to join. I'm headed home that way anyway."

Caasi coughs again. And the driver asks, "Long Covid?"

Sarai nods, yes.

"I've got three of those at home." The Truck Driver admits, then he adds, "Fucking mutating virus."

CUT TO

LATER

INTERIOR DUMP TRUCK CAB-DAWN

The low-tech exterior of the truck is juxtaposed with a high-tech interior, made up of monitors, cameras and a central climate control panel. Caasi's WHEEZING is masked by sound of BROKEN GLASS JIGGLING in the truck's rear. The Truck Driver grabs a red plaid blanket from behind his seat, puts it on Caasi's lap, then adjusts the AC and opens his glove box to grab a small aerosol can.

"This might help?" He says, as he hands Caasi an aerosol canister with a jerry-rigged face mask attached to it. "…I make these for my own sons. but only if it's ok with your mom?

"She's not my ..." Caasi starts to cough again.

"Her parents died of Covid." Sarai says, interrupting.

Caasi smiles at Sarai and whispers, "...but… I wish… she was."

"What's in it?" Sarai asks the driver.

"Just water vapor, but it does the trick."

Sarai smiles. "Ok."

"Spray it once or twice," the driver says, "and breathe in."

Then after a pause he adds, "Terrible, terrible virus."

Within a few minutes Caasi's cough quiets.

The Truck Driver motions to the radio. "You mind?"

"Of course not."

The driver fiddles with the radio. Sarai looks out the window, as the sun rises, and onto the vacant streets of Manhattan. "This new nocturnal reality is scary," Sarai says.

Still fiddling with the radio, the Driver says, "Well, hopefully come winter we can settle back into semi normal"

He stops on a RADIO TALK SHOW.

"...selfishness," a male voice on the radio says, "…may have consumed the American dream, leaving us an isolated superpower but to blame the people?"

"You mean render our arrogance, finally impotent?" a female voice replies, "'We the People' let it happen through apathy and uneducated naivety."

"That's Bullshit!" the Truck Driver blurts out. Then he looks over at Caasi and says "My apologies young man."

"...But if Technology played a role in our undoing," the male voice on the radio says, "then couldn't it also help design a way out?"

Dump Truck Driver laughs out loud.

"Look," the female voice says, "there is no tech that can save us from the Toxic Soup this last administration ushered in."

"Amen" mumbles the driver.

Sarai perks up, looks over at him, and says, "I thought you..." Sarai points to a faded and ripped sticker, on his glove box, that reads:

TRUMP

2024

SAVE AMERICA AGAIN!

"Not after what happened," the truck driver says.

"...We have farm land owned by a select few," the female voice on the radio says, "We only eat processed food, we have no rights, we have more privately owned nuclear power plants to feed the artificial intelligence machine, than schools to feed our children's minds and fresh water is being made because those thousands of new data centers and their cloud servers require more fresh water than our people do. We are being watched all the time, our moves monitored, our property taken from us, at a whim, and new infrastructure now has Chinese names on them, for God's sake. Not to mention the recent heat increase. If this isn't the end," the female guest says, "I don't know what is."

Tone deaf and evasive the male host laughs uncomfortably and says, "…but I have to admit, the whole nocturnal thing kind of suits me personally."

TIME CUT

EXTERIOR DUMP TRUCK CAB-DAWN

"POWER OF THE WOMEN OF THE MORNING SHIFT", by Oum Shatt

Alongside other construction vehicles, the dump truck moves in and out of traffic, as it crosses the 59th street bridge. The sun, directly in front of them lights up the landscape of structures half under water in the East River.

CUT TO
CONTINOUS

INTERIOR DUMP TRUCK CAB-DAWN

Sarai sits up in her seat to look out the driver's side window.

"That used to be Roosevelt Island," the driver says.

"I know," Sarai says, to herself. "Unreal."

TIME CUT

INTERIOR DUMP TRUCK CAB-DAWN

Many of the roads are marked flooded on the display monitor in Queens/Howard Beach area.

"Do people live here?" Caasi asks, pointing at the screen.

"Not much...anymore." the Truck Driver replies. "Everything here is under water most days."

TIME CUT

EXTERIOR DUMP TRUCK CAB- DAY

The truck takes an exit, onto the JFK airport causeway.

TIME CUT

INTERIOR DUMP TRUCK CAB- DAY

A modern, cast concrete and glass, bird like structure comes into view.

"What terminal?"

"Just in front of Terminal Five," Sarai points in the direction of a modern one and a half story concrete structure tucked in front, "The TWA Hotel."

"Over there?" Caasi asks. "Really? Wow."

"Not sure I can get you there with this rig, but I'll see how close I can get."

TIME CUT

The pavement, in the hot morning sun appears wavy. The building itself is fifty yards away when the truck comes to a stop.

Before Sarai opens the door, the driver says, "I hope you find what you're looking for?"

Sarai smiles, then steps out of the cab and turns for Caasi.

Still inside the truck's cab, Caasi hands the driver back his makeshift water canister.

"That's for you to keep, son."

"Really? Thank you."

Caasi turns to jump into Sarai's arms.

From outside Sarai says, "So very sorry about your seat."

Laughing, the Truck Driver says, "Water never hurt anyone."

CUT TO

INTERIOR TWA HOTEL, VESTIBULE- DAY

Sarai and Caasi walk through the double doors of the hotel,

through an insulated vestibule. Sarai notices, cameras overhead, immediately looks down, grabs two masks from a nearby dispenser, and hands one to Caasi.

"Keep this on?"

CUT TO
CONTINUOUS

INTERIOR TWA HOTEL LOBBY- DAY

CUE SONG: "SURFIN'USA" by the Beach Boys

The hotel is empty, a Beach Boys song is amplified over the hotel loud speakers as Sarai and Caasi approach a curvilinear reception counter with five unmanned rectilinear digital kiosks, that look more like 1990's hard drives with a skinny opening, like a disk drive, at the bottom and a LCD touch screen on either side. A free-standing, a 1960's Solari Board, in the middle of the reception desk. It displays fictional and ever-changing flight numbers, times and destinations while produces an ENDLESS SERIES OF MECHANICAL CLICKS.

Mesmerized by the twenty-foot-tall mechanical device hoovering above them, Caasi says, "It looks like a waterfall."

"It does," Sarai says, looking up, "doesn't it?" But Sarah quickly refocuses on the kiosks prompts, trying to check in.

"I mean, I've never seen one in person," Caasi admits, "but..."

Out of nowhere a SECURITY OFFICER walks toward them. "You can't loiter in here," he calls out. "Step back outside." He physically pushes Sarai back toward the vestibule. "You all think you can..."

Sarai pushes him back. "Don't you dare," she says.

"Stop fucking with me lady."

"I have a reservation," Sarah snaps.

"Yeah right."

Sarah points to the screen, which reads:

Sarah Abraham

Prepaid Reservation Confirmed

The officer stands down and steps away.

CUT TO

FLASHBACK

EXTERIOR SOUTH STREET SEAPORT, DOWNTOWN, NY – DAY

2020, TWELVE YEARS EARLIER

Sarai walks at the edge of a crowd of people. She is leading and holding the hand of NICOLA CUORA, her thirty something husband, who is coughing and not steady on his feet. AMBULANCE SIRENS are coming and going, from every direction.

"I can't…I need to sit down," Nicola says.

"We're almost there," Sarai says, as she stops beside a parked refrigerated eighteen-wheeler truck. A fire engine red sign with white letters on it, that runs the length of a non-descript concrete building reads:

NewYork-Presbyterian™

Lower Manhattan Hospital

comes into view as they walk past the parked truck. Beside them, two men, both wearing hazmat suits, load a black body bag onto the truck.

"Only a few more minutes," Sarai says as she pushes through a growing crowd and heading straight toward the hospital emergency entrance.

A SECURITY OFFICER stops her with his arms. "No more room for walk-ins."

Nicola starts to cough again. "My love…" he says. "I'm ok…I

just need to lay down…Let's go home." His cough is deep and raspy.

"You are not ok!" Sarai says, then turning to the officer, adds, "He needs help, please."

"Everyone here needs help lady, try uptown."

"I'm not dragging him anywhere else."

CUT TO

A NURSE emerges from the hospital doors, exhausted, with her own N95 mask still on. Sarai grabs her arm as she walks past.

"Please help my husband." Sarai pleads. The exhausted nurse turns around. Sarai pulls Nicola toward her. "My name is Sarai, this is Nicola. He's only thirty-four and he's all I have. I think he has this virus."

"It's chaos in there, sweetie," the older nurse says, exhausted.

"Please," Sarai pleads, crying.

"He might be better off at home," the nurse explains, "so many people just aren't getting better."

"But he's been sick for weeks already. Please...please try to help him."

Reluctant, the nurse grabs Nicola's hand and guides him back through the doors and into the hospital. Sarai follows, but the same officer stops her.

"No family beyond this point."

FLASHBACK END

INTERIOR T. W. A. HOTEL LOBBY- DAY

The song "SURFIN'USA" by the Beach Boys plays over the hotel speakers. The vast, column free, empty hotel lobby, with its soaring cast curvilinear concrete ceiling is flooded by daylight.

A RECEPTION CLERK, in her mid 30's walks out of a frameless, hidden door, past the security guard and up to Caasi and Sarah. Caasi is shivering from the air conditioning and her wet clothes.

"You need help?" the clerk says, as she approaches.

"Yes," Sarai says, "we do have a prepaid reservation, but the machine won't let me check in without ID."

The reception clerk smiles, "Then let's get you checked in."

She scans her ID card to reactivate the kiosk display, that had locked Sarah out.

"Last name please?"

Sarai says, "Abraham," Caasi looks up at her confused.

The woman types the name into the kiosk. "Here you are," the clerk says, smiling. "I see you're with us for two nights in our Saarinen Suite."

"Our belongings and papers were stolen…I'm sure there's a note."

"No, but I do see it's all been prepaid," the clerk says, smiling at Caasi, who is visibly shivering. "Thank you both for your patience, I just need to bypass some of these prompts."

"I'm expecting a package with our new papers in it… and I do have Euros for any incidentals."

The clerk quickly clicks onto another screen. "I'm sure it's on its way. We get at least three Amazon deliveries daily."

Caasi tugs on Sarai's arm.

Ignoring Caasi for a moment, seeing the woman struggle to get through some of the kiosk's backend prompts, Sarai says, "I hope everything is okay?"

"Everything will be fine. This machine can be very bossy. We just have to show it who is really in charge," the young woman says, laughing.

"We've had quite a journey…," Sarai says.

"It looks like it." the woman says, turning toward Caasi again.

"Do you have a toilet?" Caasi asks, addressing the woman.

"The men's room is right around that corner," the woman points at a marked door near a few feet from where the security guard stands. Then she turns her attention back toward the LCD touch screen and CLICKS through a few more screen prompts.

Caasi looks at Sarai, then runs toward the toilet.

Nervous Sarai adds, “We were robbed. The new documents should be here any hour now.”

“Of course.”

The machine spits out a room card, from the small slit at the bottom of the counter top kiosk. “Can I help with anything else Mrs. Abraham?” the Clerk asks. “We do have roller skating outside each night and an airplane, we call it Connie, right beside it to explore. I’m sure your son would enjoy both. The small plane, our pride and joy, is a 1958 Lockheed Constellation turned into our cocktail bar. You can explore it before the bar opens at 1am.”

“Do you have a restaurant here?” Sarai asks. “I know sh…. he’s hungry too.”

“No restaurants anymore, unfortunately, but you’ll find what you need in the vending machines on each floor. Just use your room key.”

CUT TO

LATER

INTERIOR T. W. A. SAARINEN SUITE – DAY

Heavy, insulated, blackout curtains, cover the floor to ceiling glass hotel room wall as daylight floods around its edges. Caasi is on the bed in the darkened room, newly showered wearing an adult sized bathrobe and talking to herself with her doll. The only artificial light in the room comes from the bathroom, where a SHOWER IS RUNNING when a KNOCK at the door startles Caasi.

The shower stops suddenly.

“Did someone just knock?” Sarai calls out.

“Yes, want me to open it?”

"Yes, tell them to wait one minute."

Caasi gets up to walk toward the door when Sarai comes out of the bathroom ahead of her, with nothing but a towel on, still dripping wet and carrying a bundle of wet clothes wrapped inside another towel.

Through the door crack, she passes the bundle off and says, "As soon as possible, please."

TIME CUT

Sarai and Caasi are asleep, snuggled beside each other. A KNOCK at the door wakes Sarai. Naked, she gets up, wraps a dry towel around herself and finds their washed and dried clothing, in a wrapped paper package on the ground.

TIME CUT

INTERIOR TWA HOTEL, HALLWAY - NIGHT

Coming out of her room, Sarai, limps down the red carpeted hallway, barefoot, but dressed in her newly washed clothing. Near the elevators a series of vending machines offer different products and food stuff. She walks toward one displaying various toiletries. With her room key in hand, she buys two pairs of flip flops, an extra small and a medium.

TIME CUT

EXTERIOR T. W. A. HOTEL, ROLLERSKATING RINK – NIGHT

A lightning storm flickers over the ocean. With the smell of jet fuel lingering in the moist air. Six adults skate on the small open-air rink. Caasi is the only child skating. She is also the only one wearing a surgical face mask.

Skating close to the wall Caasi stops and calls out to Sarai. "I wish you could skate too?"

Sarai, who is also wearing her face mask, sits on a bench beside a roller skate vending machine. "Me too," but points at her twisted ankle, makes a twisted funny face and says, "So sorry."

Caasi starts to skate again, when Sarai's attention turns to the loose shoes under the bench. Sarai, shakes off a flip flop and glides her bare foot into a FLY LONDON YOZA RED WEDGE shoe. She looks at the very feminine shoe on her foot, a perfect fit then quickly shakes it off, when she notices a couple exiting the rink.

Caasi rolls over holding onto the side wall again. "I wish the others could be here. They would think this was so much fun."

TIME CUT

INTERIOR T.W.A. HOTEL, RED LOUNGE- NIGHT

CUE SONG: "TEN LITTLE INDIANS" by the Beach Boys

Sarai and Caasi, enter the structure from the outside, as the hotel speakers play a never-ending stream of Beach Boys songs. As they walk past a sunken lounge, carpeted in red, filled with matching red upholstered Knoll Saarinen Tulip Chairs, scattered with the white dots of Knoll's Saarinen Tulip Tables they pass many other people, masked and unmasked, who transit, with and without luggage, through the hotel.

CUT TO
CONTINUOUS

INTERIOR T.W.A. HOTEL, RECEPTION- NIGHT

Two Asian children, an older boy and his sister, stand near the vintage red Fiat on display, near the hotel entrance. Excited to see a girl her age, Caasi runs toward them.

"It seems too small to be real." Caasi says, smiling.

The siblings look at her, then speak Chinese with each other and run away.

CUT TO

Sarai stands a few feet away, at a concierge counter, speaking to a young man in his thirties.

"...we do get multiple Amazon deliveries a day."

Caasi walks up to Sarai and stands beside her silently.

"What times do they come?" Sarai asks, "I'm waiting for an important package."

"The next one will come at midnight. We can alert you, if you like?

"Yes please."

After a pause Sarai asks, "Just curious, is there any way to extend our stay, on the existing card, if we need?"

"We would need to see if we have any vacancies first, then run your credit card again. Would you like me to look?"

LATER

INTERIOR T.W.A. HOTEL, ELEVATOR / VESTIBULE – NIGHT

Tucked around the corner of the lounge, Sarai and Caasi wait for an elevator in a carpeted vestibule.

"Do you think it will come in time? Caasi asks, concerned.

"Please don't think about it. Of course it will." After a short pause, Sarah adds, "Let's try and have fun."

CUT TO

CONTINUOUS

Stepping into the elevator Caasi rips her facemask off.

Looking up at a camera, Sarai says, "Wait until we get to the room Caasi."

"Ok."

Then Caasi points to a button that reads:

ROOFTOP POOL

“Can we?”

“But it just started raining?”

“I just want to see. Could we please…This really is very fun.”

CUT TO

INTERIOR T.W.A. HOTEL, ROOFTOP VESTIBULE - NIGHT

The elevator door opens and Caasi runs ahead into a small brightly lit, tiled space. She runs over an inlayed gold T.W.A. emblem that is imbedded in the terrazzo floor. Caasi is stopped in her enthusiasm by a locked glass door. She pushes her face up against the glass trying to see into the darkness outside. The sound of machinery HUMS through the glazing.

“I can’t see anything.”

Sarai walks up behind her and looks too.

“Where’s the pool?” Caasi asks.

Sarai cups her hands onto the glass to look out.

CUT TO

CONTINUOUS

EXTERIOR T.W.A. HOTEL, ROOFTOP POOL - NIGHT

A flash of lightning lights four EPSILON Atmospheric Water Generators. Their milky white plastic containers sit filled with water in the empty pool.

Just beyond the roof deck, 747 airplanes, six stories below, are taxiing out of their gates and onto a nearby runway.

CUT TO

CONTINUOUS

INTERIOR T.W.A. HOTEL, ROOFTOP VESTIBULE - NIGHT

“What do they do?” Caasi asks, still looking through the glass door.

"The machines?" Sarai asks

"Aren't they the same ones we saw under the buildings?"

"Yeah," Sarai says. "Weird."

"Why weird?"

"They make fresh water from the moisture in the air."

"To drink? Really? That's cool."

"But," Sarai says, "I thought they were only being used to cool the data centers." Then mostly to herself, she says, "I guess I had no idea how widely used this technology was."

TIME CUT

INTERIOR T.W.A. SAARINEN SUITE – NIGHT

Inside their SILENT unlit room, Caasi sits on the floor, in front of the floor to ceiling glass wall, watching planes exit their gates and maneuver toward the runway, while others take off.

Caasi startles when the room door OPENS. She turns around for a second, but looks back outside. "We're going on one of these big airplanes?"

"Yes," Sarai says, unpacking a small bag of food onto the desk.

Caasi whispers, to herself, "How do they stay in the sky? I've never been on one before. I know they stay in the sky, but it's just crazy to think about." Then Caasi turns around. "Did the tickets come yet?"

"No, not yet, but I got you some food, come eat."

CUT TO

INTERIOR T.W.A. HOTEL- NIGHT

CUE SONG: "SUMMERTIME BLUES" by The Beach Boys

MONTAGE

The full length, mechanical, bedroom blinds open automatically, with a MECANICAL HUM Sarai and Caasi wake from the noise.

CUT TO

Sarai is at the reception desk speaking to another clerk. This time, an older female. Caasi plays make believe, a few feet away, sitting in the small red car.

CUT TO

Sarai and Caasi, sit in various spots on the red furniture in the lounge.

CUT TO

Sarai and Caasi explore the full-size plane, named Connie, outside. They can see the roller rink outside the window.

CUT TO

Caasi roller skates, as Sarai watches from the bench.

CUT TO

Back at reception, Sarai asks about the package, while Caasi looks up at the Solari Board, and is mesmerized by its rhythmic mechanical CLICKING.

MONTAGE END

TIME CUT

INTERIOR T.W.A. HOTEL, HALLWAY – NIGHT

Sarai exits their room, barefoot, and limps down the red carpeted hallway toward the vending machines.

CUT TO

Sarai inserts her room card in the vending machine titled PARIS CAFÉ. While studying the strange display of prepackaged foods.

"Nothing French," she says to herself. "So odd."

An elevator door opens and a PORTER exits, and walks past Sarai.

"Room 823?" Sarai calls out, on a whim.

The Porter turns around, surprised. "Yes, something's at the front desk for you. We tried calling, but didn't get an answer." He walks back to Sarai and hands her a small paper ticket.

CUT TO

INTERIOR T.W.A. SAARINEN SUITE – NIGHT

Caasi sits on the edge of the bed, wearing an adult sized bathrobe, watching TV.

An INTERVIEWEE on the television says, "...we interfered with our atmosphere with the refrigerant Freon. It depleted the ozone.

The REPORTER asks, "So, you're saying this is our fault?"

"With chemicals like black market Freon still being bought and sold in the U.S. Yes, it might be?"

"But it was banned in 2020."

"On paper, but there's still a pretty robust market for it in the states."

The hotel door opens. Caasi is surprised and turns the TV off, as if caught doing something wrong.

"I'll be right back...," Sarai says, slipping into her flip flops by the door, "I think it's here."

The door SLAMS shut.

Relieved, Caasi turns the TV back on, but the door opens a second later.

"Try not to fill your mind with bad news," Sarai calls out from the cracked doorway.

The door SLAMS SHUT again.

TIME CUT

Caasi sits back, with her feet up in red upholstered Knoll Eero Saarinen Womb Chair, still watching the same TV program. The controller is beside her on the arm.

"...The natural resources we have on this planet, we'll always have" the voice on the television says. "Because they are finite.

Nothing comes into or out of the closed loop. So, since water can't be made, it is stolen from somewhere else always."

The hotel DOOR OPENS.

Caasi clicks the TV off and jumps up, feeling guilty and runs to greet Sarai.

Sarai has a padded manilla envelope in one hand and a bag of vending machine food in the other.

"Television is so bad for you." Sarai says.

"I know. I know. I'm sorry. But I was watching some science stuff. Some of it made sense."

"Just be careful," Sarai says, handing Caasi the bag of food. "I'm sorry that took so long,"

While Caasi rummages through the prepackaged foods Sarai opens the envelope and dumps its small contents onto the bed; Two passports, a new phone and a charger.

Sarai shakes the envelope more, then looks inside it.

Caasi watches her. "What are you looking for?"

"A note?"

"From Mr. Abraham?" Caasi asks, stuffing a Twinkie in her mouth.

TIME CUT

Sarai plugs the new phone into the wall to charge, while Caasi studies her new passport. "How'd I get this name?"

Sarai gets up off the floor. "Caasi, Twinkie's are not food. I got that for you to eat after." Then she looks at Caasi's passport, pauses for a moment and says, "It's your name... but backwards."

Caasi takes another bite of her Twinkie. "Isaac Abraham. Weird. I like Caasi Aditi better."

"Please eat something real."

Caasi laughs, when she looks at Sarai's passport, which reads:

"He spelled your first name wrong?

Sarai looks at it. "He did…didn't he."

"But he made you, my mom, so…"

"…and kind of his wife apparently," Sarai adds, interrupting.

"I'm going to call you 'Mamma Sarah' from now on," Caasi says laughing, while rummaging through the bag of prepackaged food again.

TIME CUT

INTERIOR T.W.A. SAARINEN SUITE – NIGHT

CUE SONG: "HIGH ENOUGH" by K Flay

The mechanical blinds are drawn, despite it still being dark outside. Caasi is asleep and the only light on is behind the closed bathroom door where the TOILET IS RUNNING.

Inside the bathroom, Sarah is wearing a white bathrobe and sits on the closed toilet lid, jiggling the toilet handle.

"Stupid waste," she whispers under her breath. Then she continues to text Abraham on her new phone:

You're such a surprise. You've never been so sexually explicit before.

Instantly, Abraham texts back:

Oh babe, you know I always want you.

Caught off guard by his uncharacteristic sexual demeanor, Sarah waits a few seconds to respond then texts:

Please share why it took so long for you to reply? It's so unlike you.

Oh babe.

Abraham replies. Sarah stares at the new salutation, not sure what to make of it. Then Abraham continues:

It was the new number that threw me. You have no idea how many calls I get in a day.

Annoyed that the toilet doesn't stop running, Sarah jiggles the toilet handle again. Abraham texts, a third time in a row:

I'll get rid of your other number and replace it. Ok?

Sarah finally replies:

The last few days were so hard. Maybe this is just a fever dream? Please tell me it's not.

I'm sure it's more.

Abraham replies, then he adds:

I'm not letting you leave tonight.

CUT TO

INTERIOR T.W.A. CORRIDOR TO TERMINAL – NIGHT

CUE SONG: "I GET AROUND" by The Beach Boys

Muffled AIRPORT ANNOUNCEMENTS come from a set of double doors at the end of a long corridor. Sarah and Caasi hold hands as they walk down the red carpeted, half circle corridor, which is on a slight incline, toward terminal 5's entrance.

CUT TO

INTERIOR JFK INTERNATIONAL AIRPORT, TERMINAL – DAY

A RINGTONE sounds over the airport speaker followed by an ANNOUNCEMENT as Caasi and Sarah enter the chaotic terminal.

"You are Isaac now," Sarai whispers. "Do not answer to any other name. We need to get through this and onto that plane without being stopped."

As Sarah navigates the new environment, Caasi stays close by, but looks around eagerly.

The AIRPORT ANNOUNCEMENTS are incessant, loud and mostly unintelligible.

Sarah locates the self-serve kiosks for Emirates Airline, and begins the check-in process, but a prompt on the screen flashes:

Enter Proof of Covid31 Vaccine

"Oh shit, our vaccine data."

Caasi looks at Sarah. "Is that bad?"

"I forgot to bring it." Sarah says, looking around for help.

"Is it bad, Sarah?" Caasi says, a bit louder, getting anxious.

"Shhh," Sarah whispers, kissing Caasi's forehead, "Only call me mamma in here. Ok?"

Sarah flags down an AIRLINE EMPLOYEE, a man in his forties.

"What does this mean?" she asks, in an Italian accent, pointing to the screen.

He walks over to take a look. "It means, if you're an American citizen and you haven't had the vaccine yet, you don't fly. Not our rules, the rest of the world's."

"Ma, we are Italiano…a Roma?" Sarah continues, in broken English with an Italian accent.

"Then, you need proof of a negative Rapid Antigen Test before, you can check- in." The man points to a long line of white testing cubes, at the edge of the cavernous space with a large sign above them, in seven languages, that reads:

IF YOU DON'T PASS, YOU DON'T TRAVEL

WER NICHT BESTEHT, REIST NICHT

SI VOUS NE RÉUSSISSEZ PAS, VOUS NE VOYAGEZ PAS

SE NON PASSI, NON VIAGGI

若您未通過考試，請勿出行

നിങ്ങൾ പാസ്സാക്കിയില്ലെങ്കിൽ നിങ്ങൾ യാത്ര ചെയ്യരുത്

ЕСЛИ ВЫ НЕ ПРОЙДЕТЕ, ВЫ НЕ ПУТЕШЕСТВУЕТЕ

"Get a test," the airline employee says, "The QR code you get, will start the check-in process."

CUT TO

INTERIOR JFK TESTING CUBE- NIGHT

Sarah watches large monitors above the security checkpoint that display names and faces of US citizens, as she and Caasi wait on a testing line.

Every few seconds a new set of pictures flash.

"Who are they all?" Caasi whispers.

"People in trouble… that got away."

A female MEDICAL TECHNICIAN, in her twenties, approaches speaking in an indifferent tone, "Fore finger…it's just a small prick."

Sarah and Caasi hold out their fingers. Caasi winces and looks away.

The Med Tech takes a drop of blood from each. Then pricks another TRAVELER, a young man in his twenties, who wears a bright blue baseball cap with a white letter, T outlined in red, on it.

He looks straight at Caasi when he shouts, in a loud Texan accent "WHOO-WEE! It's always the tiniest pricks that cause the most pain."

The Med Tech holds up a digital QR code on a small iPad for Sarah to scan with her phone, then another for the Texan.

"Once you get your results," the Med Tech says, "use this code to check-in."

A younger picture of Caasi flashes up on the huge monitor yards away. Sarah immediately turns Caasi around into her belly.

"What happened?" Caasi asks.

"Just keep your mask on and do not take it off."

Caasi looks up at Sarah. "Am I in trouble?"

Sarah turns to the Texan and in her Italian accent, asks, "How do you say? Capello?" as she points to the man's baseball cap.

"De-what?" he says, in his own heavy southern accent, "My hat?"

Sarah puts her hand on Caasi's head and says, "Si, mio figlio." to indicate her son, then she says, "He likes the team very much."

The man smiles, takes off his hat and looks at it. "I had no idea baseball was big in Italy?" Then to Caasi, he says, "You want a souvenir little buddy?" He hands Caasi the Royal Blue hat. "It's not the Mets, but the Texas Rangers are molto bene."

He laughs at his attempt at Italian, just as Sarah's phone PINGS twice, with their two test results;

NEGATIVE S. Abraham

NEGATIVE I. Abraham

"Grazie mille," Sarah says, to the young man, encouraging Caasi to accept the gift. Confused, Caasi reluctantly accepts.

Sarah leans down, and whispers, "Please put that hat on and keep your head down."

The man's phone PINGS he instantly walks back toward the Med Tech. In a booming voice he asks, "What do you mean positive? I got your damn vaccine a few days ago."

TIME CUT

INTERIOR JFK AIRPORT SECURITY- NIGHT

Sarah pulls her coat's hood up, as she and Caasi get onto the Foreigner security line. There are hundreds of men, women and children moving through the space. A US military presence, in the form of humanoid robots, stand between the foreign security line and the US citizen line. The United States citizens are more heavily scrutinized, forced to unlock their phones and laptops and hand them over to the few humans working on the line with devices of their own to scan the device's contents.

CUT TO

Twenty 'Clear' Kiosks scan the United States citizen's retinas and finger prints. Sarah's line, for the EU, is without scans and moves quickly.

Caasi whispers, "Why are they doing all of that?"

"This is what an authoritarian government looks like."

Just then a scuffle on the US line breaks, a few feet behind them.

"What happened to our rights?" a middle-aged American traveler, screams as he's pulled aside by two, armed, human military officers. "My child support is none of your business."

Distracted by the commotion, A SECURITY OFFICER, in his mid-forties, who sits behind a plexiglass shield, barely looks at Sarah's two passports before waving them through. Sarah and Caasi continue toward the luggage and body scan.

"Keep your hat on until I say so."

Caasi nods.

The plastic bins the other travelers fill with all of their carry-on luggage are overloaded, but Sarah's. She takes her small back off then her long coat.

"Give me your flip flops, baby," she tells Caasi.

"But… my bare feet?"

"I'm sorry our shoes never dried out. We'll buy new shoes once we're inside."

Sarah places Caasi's small flip flops into the bin, then two red wedge London Fly shoes on top, shoes she stole from beside the roller rink.

CUT TO

INTERIOR 747 – DAY

A beautiful European STEWARDESS wearing an Emirates red pill box hat and scarf walks down the aisle. Caasi is already sitting, buckled into her window seat. Her face is pressed up against the rounded window. Sarah turns her new phone onto airplane mode, then leans into Caasi and looks out of the window too.

CUT TO

EXTERIOR EMIRATES 747 AIRPLANE – DAY

The plane taxis down the runway as the sun rises.

CUT TO

INTERIOR 747 AIRPLANE – DAY

Sarah closes her eyes, taking deep breaths and under her breath she says, "The only things worth doing in life require a countdown."

"Look, look," Caasi says, tugging on Sarah's arm.

Out of the window, both watch as ocean water jumps a newly built seawall on the edge of the runway.

"I had no idea how close to the ocean we were," Caasi says

TANYA H. VAN COTT

Part II

Data Transfer [GREEN]

Roma

“This invasion of one's mind by ready-made phrases can only be prevented if one is constantly on guard against them…”

Politics and the English Language

by George Orwell 1945

TANYA H. VAN COTT

INTERIOR ROME AIRPORT – DAY

Inside Rome's Fiumicino Airport, Sarah and Caasi are surrounded by other passengers dragging their luggage toward the airport exit. Sarah scans the line of glass exit doors, then stands on her toes to look for Abraham, through the sea of movement.

"Do you see him?" Caasi asks.

"No."

"Do you think he still looks like the picture you showed me?"

Sarah pauses, looks at Caasi, then laughs uncomfortably, "…I hope so?"

"Maybe he's somewhere else?"

"Ok you're right, let's go back to baggage claim, then"

"But we don't have a bag?"

"Well, he doesn't know that? I'm sure he can't imagine what we left behind in order to get here."

CUT TO

INTERIOR ROME AIRPORT, BAGGAGE CLAIM- DAY

Sarah and Cassi back track, walking against the flow of pedestrian traffic, past a line of empty baggage carousels and back to their posted flight. Sarah suddenly pulls Caasi toward a restroom after noticing a man following them. "Hold my hand and come with me."

"Are you feeling, ok?"

"Yes, Shhh."

Sarah and Caasi slip into the restroom. After a few seconds Sarah's peers around a corner and sees the same man waiting. A large group of East Asian women and children exit the bathroom together and Sarah walks into the center of the group with Caasi.

Within a few feet, she hears the same man call out, "Sarah?"

Sarah speeds up, moving ahead of the tourists.

Caasi looks back and notices the man, "Who is that?

Sarah picks up her pace, "I have no idea."

The man runs up behind them, then grabs Sarah's arm and in an Italian accent he says, "Sarah?"

Sarah shakes him off, defensive.

"I'm Abraham's friend, Luigi. He sent me to pick you both up."

Sarah freezes for a minute. Caasi stands behind Sarah hiding.

Happy to have found her, Luigi, lunges forward to kiss Sarah on both cheeks, then with a big smile reaches his hand out to Caasi.

"My name is Luigi."

"I'm...," Caasi looks at Sarah, "Isaac."

In broken English, Luigi says, "Come. We go now. You must be Molto Stanco." Luigi makes a gesture with his hands about sleep. "No...how do you say? Bagagli?" he adds.

"Luggage? Sarah says, "No. Questo e."

"Ahh," Luigi says, in Italian, "you speak Italian?"

"It's been a while." Sarah says.

"Va bene!"

Luigi leads them both out of the airport doors.

CUT TO

LATER

INTERIOR LUIGI'S CAR, EN ROUTE TO ROME - DAY

With the stick shift in his hand, Luigi races the other cars en route to Roma Centro. Caasi, sits in the back seat and looks out the window at the strange brown landscape and even stranger Umbrella Pine evergreens.

"Will Abraham be...," Sarah starts to ask.

"No, no," Luigi says, in Italian, looking straight ahead driving aggressively through traffic. "He is still traveling."

In Italian, Sarah asks, "Do you work for him?"

Luigi laughs. "No, his business partner actually."

"Also, a chaos engineer, then?"

"No, Bella. A hydrologist."

"He never mentioned you?"

"Never?" Luigi laughs again. Then in English, he says, "Abraham doesn't how do you say… share easily."

TIME CUT

As the car speeds past the Vatican Sarah receives a text from Abraham:

…I told you last week?

Sarah texts back:

Did you?

Abraham says:

Babe, you know I'm pulled in a lot of directions?

Sarah does not reply, shocked by the change in demeanor from the night before.

After a few minutes Abraham sends another text:

Luigi is my better half. He will take care of you. Glad you landed safely.

CUT TO

EXTERIOR PONTE SISTO BRIDGE, TRASTEVERE- DAY

Luigi's car pulls up beside the Tiber River, at Piazza Trilussa.

The Tiber river's GREEN WATER, a few yards away is breaching the river's embankment.

CONTINUOUS

INTERIOR LUIGI'S CAR, TRASTERVERE – DAY

"That's not normal," Sarah says, looking out of the window. "I remember the Tiber being at least ten to fifteen feet lower.

"The full moon," Luigi says, in Italian, but then in English he adds, "No, it is not normale, you are right."

CUT TO

EXTERIOR LUIGI'S CAR, TRASTERVERE – DAY

As Sarah steps out of the car with Caasi following her lead. Luigi, hands Sarah keys through the open passenger window and points toward a 17th century weathered three-story stucco building tucked in-between other buildings just like it.

"I need to go park," he calls out of the car. "Numero Settantanove, the apartamento is on the secondo piano."

After Luigi drives off, Sarah and Caasi start to walk across the cobble stone piazza, filled with parked cars when an Ambulance, with its WAIL AND YELP sound passes by, flashing its blue lights.

CUT TO

INTERIOR ABRAHAM'S APT.- DAY

As Sarah unlocks the front door of the apartment, Caasi enters the unlit and SILENT space. Sarah turns a hallway light on with an audible CLICK. And Caasi runs ahead, exploring.

Out of sight Caasi calls out, "I wonder which one is ours?"

"We'll have to wait and see." Sarah says as she looks into the first room off of the foyer. It's a small library with books on opposite walls framing one shuttered window in the center.

Caasi pokes her head into the hallway from a different room. “Well, I hope we get this one.”

Still standing only feet from the vestibule, Sarah studies the room’s contents, books, by Kurt Vonnegut, Nabokov, Nicholson Baker, Anais Nin. Also books on history, geography, topography, geology and hydrology.

Sarah smiles to herself.

“Come see!” Caasi calls out.

Sarah walks toward Caasi, past a small bedroom, a small kitchen and a bathroom all hanging off or the one narrow corridor. Each room is painted the same hospital green except for the largest room at the end of the hallway, it’s painted a bright white. Caasi is already laying down, on the crisp white sheets of a newly made queen size bed. A large green Persian area rug covers the amber colored floor tiles.

Sarah walks into the room, drops her backpack and coat onto the bed, then heads over to a desk in front of a large shuttered window.

From behind Sarah, Caasi asks, “How long will we be here?”

Without turning around, Sarah says, “For a while.”

“Look, we can put your light here,” Caasi says.

Sarah turns to see Caasi already unpacking the small teepee light, places it on a low stone slab.

Two computer monitors, on the desk, glow and dance in sleep mode. Sarah looks at through a large set of construction print outs of open in front of them. They are a variety of city plans. Every piazza has a red dot marked in the center of it with a series of dashed lines leading from it.

“I like it,” Caasi adds, “Do you think they’ll be other kids?”

Sarah gently flips through the set of drawings. There are red dots in every piazza of a variety of Italian cities and town plans.

"Not sure?"

The front door OPENS, then SLAMS shut.

Out of breath, and still out of site, Luigi says, mostly to himself, "Tanti ancora guidano a quest'ora…Crazy." Then, with sweat still on his forehead, he appears in the doorway of the large bedroom.

"Trastevere will never be easy," he says, in English, turning toward Caasi, he continues, "I see you are making yourself...how do you say? …Comoda?"

Sarah turns and says, "Comfortable?"

"Va bene!"

"Mr. Luigi," Caasi asks, from the floor, already under the bed, with the light's electric cord in her hand, "This plug doesn't work."

"Perche, you need a, come dici?"

Sarah interrupts, "Adaptor?"

"Is this one our room?" Caasi asks.

"You like my room, little Isaac?"

Sarah asks confused. "Isn't this Abraham's apartment?"

"Of course, his building in fact," Luigi says, in Italian. " But he travels so much…" Luigi walks toward Sarah, then collects a book off his desk. "Oh, it's a bit complicated Bella," he says, in English, " My apartamento is one flight up, but is occupato." Then he turns to Caasi and says, "I can fix up the bibliotheca for you, and Signorina Sarah, can have this one?"

Luigi sees Caasi's disappointed face then he says, "But, yes, if you like this one it is yours to share…until Abraham returns."

"And when will that be?" Sarah asks, walking away from the desk.

"I never really know."

Spotting the desk chair, Caasi runs across the room, past Sarah, and plops down onto the leaf green upholstered Knoll Morrison Hannah desk chair, with its polished aluminum frame.

"Caasi...," Sarah says, admonishing her.

"But…we have the same happy smiling chair in New York. Ours… I mean yours is red and black, though."

"A popular chair, a chameleon of sorts because of its many colors," Luigi says, sitting down on another chair, a HERMAN MILLER Eames Lounge chair, in white leather. "Mobilia is meant to be sat on and prodotti…," he says, bending over to turn a record player, beside it on, "… to be used."

CUT TO

LATER

INTERIOR ABRAHAM'S APT.- NIGHT

Sarah wakes in the large bed to the sound of Luigi chatting, in English with Caasi in the kitchen.

"The bed is comfy? No?" he asks.

"Yes, thank you." Caasi says.

CUT TO

CONTINUOUS

In the kitchen Luigi sits at the table, facing the hallway door, and across from Caasi.

"I can still make the front room ready for you. But when Abraham arrives..." Luigi stops just as Sarah appears standing in the kitchen doorway.

Caasi shakes her head no, but then turns around.

Sarah nudges Caasi then Caasi reluctantly says, "Ok...thank you."

The sound of CHILDREN LAUGHING comes from the apartment above. This changes Caasi's demeanor instantly. She looks up at the ceiling, surprised.

"I will introduce you... after our meal," Luigi says. "Va bene?"

Caasi smiles and shakes her head, yes.

CUT TO

LATER

EXTERIOR TRASTEVERE, ROME- NIGHT

Luigi, Sarah, Caasi and two younger girls, sisters, walk through the streets outside of the apartment in Trastevere. Luigi carries the youngest, FLORI, a three-year-old with dark hair in pig tails, on his shoulders. Caasi walks beside Luigi, but holds hands with, LILIYA, a nine-year-old with wavy, strawberry blonde hair.

Luigi speaks in Italian finishing an animated children's story for the younger girls.

They all walk slowly because of Sarah's twisted ankle.

Sarah translates the Italian story for Caasi. "…Luigi said, 'Every night, all night long a ghost bakes fresh bread for everyone. The only proof he exists, is the smell of fresh bread in the streets and the flour footsteps he's left behind."

"Is that true?" Caasi asks.

Luigi taps Caasi's shoulder and points to the white flour footsteps, on the cobblestone in front of them. "How else?" he says, in English.

Caasi laughs.

Liliya pulls on Caasi to run ahead. They disappear around a corner into a large piazza.

"It's like nothing has changed in fifteen years," Sarah says.

"Italians have weathered so much history here, what could change?" Luigi says, in Italian, snuggling Flori, who is now in his arms, sucking her thumb, and leaning against his chest.

CUT TO

EXTERIOR PIAZZA SANTA MARIA – NIGHT

CUE SONG: “SANS PARADE”, by Hyperborea

Sarah, Luigi and Flori, turn the corner into the Piazza Santa Maria, in Trastevere. A twelve-foot-tall machine, with its large yellow Epsilon Logo stands where a fountain once was, in the center of the piazza.

Luigi notices Sarah’s shock.

“Is every piazza like this now?” The LOUD MECAHNICAL HUM drowns out the human sounds of the popular gathering place.“Trevi or Piazza Navonna?”

Luigi laughs, “We are Romans, not Americans.” As they walk toward the machine, locals fill bottles at the accessible spigots. Many others eat at the small outdoor restaurants.

Then after a pause Luigi says, “We’ve been working on this for years.”

“Who?”

“Our company, G.W.S.”

“I thought Abraham was involved with weather? He was never very clear.”

“Yes, everything to do with the atmosphere, precipitation and water infrastructure.”

Walking around the large machine, Sarah asks, “I thought these were only for cooling data ‘centers? Is the water crisis that bad already?”

“Si, the crisis is real, Bella.”

“In the U.S.,” Sarah says, “no one knows what to believe anymore, so no one believes anything.”

“These piazza installations are show of good will,” Luigi

explains. "Recognition that we aren't consuming their water anymore. That we will eventually replace the fresh water Ai has already consumed."

"Oh boy."

"The Paris Basin, the English Chalk Aquifer, and our own Adige Valley Aquifer, in northern Italy were struggling to keep up with demand, especially now with erratic precipitation. My team has been working with European Governments for the past two years on ways to integrate this new water generation technology into existing infrastructure as well."

"Ahhh, the drawings on your desk."

"Si."

The Basilica di Santa Maria in Trastevere, the most prominent structure in the piazza, comes into view. The doors to the church are open and the light from inside contrasts the dimly lit piazza.

Caasi and Liliya run up.

"Can we get an ice cream? I mean gelato?" Caasi annunciates the new word slowly. "Liliya showed me all these colorful pink, yellow and orange ones."

Flori perks up at the word Gelato and gets restless in Luigi's arms.

"Sorbetto!" Luigi says, smiling. "Milk has been a hard one to replace. What flavor do you like? Flori likes Limone." Luigi puts Flori down, then he asks Liliya, in Italian, "Please hold her hand tight."

"I liked all the rosa ones," Caasi says, looking at Liliya, who smiles.

"No, mango or arancia?" Luigi asks, smiling at Sarah, as he hands the older girls a few euro each.

Caasi looks at Sarah, confused.

"He's making a tech joke," Sarah says, "MANGO used to be a Big Tech acronym in the US, circa 2025, but ORANGE is the newest acronym here in the EU since the addition of Epsilon and ReitAi.

Caasi laughs, still confused, but runs toward the gelateria with Liliya.

Sarah walks toward the open doors of the Church

As she peers into its interior. Three stories of scaffolding with stairs up to each level occupy the entire interior with hundreds of men women and children living on the scaffolding.

A few seconds later, Luigi looks over Sarah's shoulder and in Italian, says, "All those dislocated from the coastal flooding."

"All relocated here, in Trastevere?"

"No, all churches, throughout Italy..." he says, in Italian. "The churches remain cool naturally because of the heavy stone architecture…" An AMBULANCE SIREN fades in then out in the near distance making the rest of what Luigi says not understandable. "…not in the United States?" Luigi asks.

Sarah laughs. "No. We do other horrendous things with displaced people. But you must already know that."

"Si. Molto triste."

Sarah turns around to look at Luigi, but notices Caasi having a breathing fit in front of the Gelateria.

Sarah runs toward Caasi, picks her up and runs next door to a restaurant with outdoor seating. Luigi and the children follow.

Sarah struggles with her Italian, while attempting to talk to a waiter.

"A wet cloth! Un panno...

"Un panno Bagnato velocemente!" Luigi interrupts, in a loud voice. Caasi lays in Sarah's arms and has tears rolling down her face as she struggles to get a full breath. The waiter returns and pours water onto a cotton napkin. Sarah puts it on Caasi's face, who instantly begins to breath a bit easier.

"Does this happen often?" Luigi asks, in Italian.

"Too often."

"Siate Happy? My Amico." Liliya says, in broken English trying to make Caasi smile.

Flori follows her sisters lead and holds Caasi's hand and smiles.

CUT TO

LATER

INTERIOR LUIGI'S APARTMENT- NIGHT

Inside Luigi's upstairs apartment TETYANA, the petite blonde, mother of Flori and Liliya looks through her bedroom armoire with Sarah. The three children, trying to communicate in English and Italian run in and out of the bedroom playing a game with a ball.

In a heavy Ukrainian accent, Tetyana says, "This one may fit?" She hands Sarah a dark green dress with small red flowers on it.

From the other end of the corridor, close to the door, Liliya says, "Prendi la palla."

"The ball?" Caasi calls back, not sure.

"Si. Palla."

Standing half naked, Sarah looks at the dress in her hands. "I haven't worn a dress in ages." After having slipped it on, Sarah takes a deep breath as she looks at herself in the long mirror attached to the wardrobe door.

"Not as loose as I hoped," says Tetyana, "but sexy."

Sarah laughs, "I thought sexy was over for me."

"Tomorrow, we go to Porta Portese Market, to see what else we can find for you both."

"I hope Liliya isn't too disappointed that Caasi isn't a boy?"

Tetyana laughs, "She still thinks she is. It's the short hair. But likes that she can share some clothing.

Sarah continues to look at herself in the mirror, tilting her head to the left than to the right. "With my short hair, and wearing a dress I hardly recognize myself anymore." Then Sarah mumbles to herself, "I hope he's not too disappointed by our age difference?"

"He'd be a fool." Tetyana says, as she spins Sarah around.

"You and Luigi are, both, so kind."

"Only when we want to be."

Sarah looks at Tetyana, confused.

"Everyone has two sides?" Tetyana says, "Don't we?"

CUT TO

LATER

INTERIOR LUIGI'S APT KITCHEN – NIGHT

Sarah wears the new green dress and sits at Teyana's kitchen table texting Abraham:

Italy has changed so much in fifteen years. It's definitely everything and nothing like I imagined.

Tetyana cuts red peppers, then gets up suddenly to pull some basil leaves off a vertical Hydroponic Garden covering one wall in the small kitchen. Sarah looks back down and continues to text:

Liliya and Flori are such a welcome surprise. I wish you could see how happy Caasi is to finally have girl playmates. It's like a switch was flipped inside her soul. I miss you and our hour-long digital chats.

The CRIES OF BABY FLORI stop both women. Caasi walks into

the kitchen holding Flori's hand, as big tears fall from the baby's eyes, "I think she's upset that Liliya doesn't want her to play with us?"

Tetyana immediately gets up and walks out of the room with Flori and Caasi trailing behind her. From down the hall, Tetyana says something to Liliya in Ukrainian followed by silence. Sarah looks back down at her phone and texts:

`You seem especially busy at the moment? I wish you mentioned where you were the other night? You used to always let me know. So sad not to see you at the airport, but surprised you never mentioned Luigi? What a lovely thoughtful man. I'm so glad you have so many kind people in your life, Abraham.`

Sarah hits send, just as an AMBULANCE SIREN, fades in and then out again. Teyana walks back into the kitchen taking her seat at the table when Sarah adds a final text:

`When we finally meet, for the very first time, which I hope is very, very soon, I cannot wait to really share our ideas, bodies and hearts.`

CUT TO

Tetyana cooks at the stove as the three children run in and out of the kitchen in a frenzy of happiness. Sarah helps set the table, leaving her phone face up on the table.

CUT TO

All five eat together in the tight kitchen laughing and chatting in half English, Ukranian and Italian. Sarah helps cut up some of Flori food.

CUT TO

The two women clean up at the sink with WATER RUNNING and chat quietly, as the three children continue to run past the kitchen, playing in the hallway when Sarah's phone, sends out a LOUD PING.

Sarah smiles. She stops to dry her hands then picks up her phone.

The smile instantly falls off of Sarah's face, as she stares at the reply. She puts the phone down on the table, face up, and walks out of the kitchen without any words. The phone shows Abraham's reply:

I see you suddenly need a lot of attention?

CUT TO
LATER

INTERIOR ABRAHAM'S APARTMENT- DAY

Downstairs, inside the bedroom, ever present AMBULANCE SIRENS fade in then out, as Caasi sleeps curled around her doll beside Sarah in the large bed. The glow of Sarah's phone lights her face, in the darkened shuttered room that has daylight pouring in through its cracks. Sarah texts:

I have so many fantasies for us.

I'm sorry if I flooded you with information earlier, I'm sure you're busy.

An AMBULANCE SIREN fades out as Sarah continues:

Such a torture and blessing this technology can be.

Earlier really hurt…

Before she hits send, on her admission of pain, a text from Abraham comes in:

I'm here. Tell me more about your fantasies?

While reading his words, she begins erasing hers. One letter at a time.

CUT TO

EXTERIOR PORTA PORTESE MARKET- NIGHT

WEEK ONE

The open-air night market is overwhelming. Hundreds of Italians shop for new and thrifted items. Luigi and Sarah walk behind Tetyana and the three girls through the bustling flea market.

"...but there's suddenly no pattern?" Sarah says. Luigi locks arms with Sarah and laughs.

"Consistently Inconsistent", says Luigi, in Italian, "the perfect title for a book about our Abraham."

"But the Jekyll and Hyde behavior is so new?" Sarah says, to Luigi.

Caasi turns to grab Sarah's hand and says, "Is love always complicated?"

Tetyana turns to look at Luigi and Sarah because the adult conversation is inappropriate. But a sudden intense downpour pushes them all under one of the vendors tents filled with brightly colored hair clips, bows, and headbands. Desperate not to get her phone wet, Sarah asks for a small plastic bag.

"Not unless you buy something," the vendor says to her, in Italian.

CUT TO

LATER

INTERIOR ABRAHAM'S APT. - NIGHT

CUE SONG: "SEXY AND I KNOW IT", by LMFAO

In the large bedroom, Luigi and the three girls dance in the middle of the room LAUGHING. Each of them has small hair clips made of colorful bows in their hair. Caasi's short bangs are pulled to the side by a small green bow. Sarah wears a new lighter green floral dress, similar to the wrap dress Tetyana gave her. Sarah sits on the bed, rereading a recent text from Abraham:

...If I am stressed, I lose focus on anything else until I resolve what is stressing me out.

Then she sends her response:

Me too, hence this text thread. I am an open book emotionally and you are suddenly in a language I am desperately trying to understand. What changed between us?

TIME CUT

The children are laying on the carpet, coloring quietly together. Sarah folds some of Caasi's new clothes on the bed, when Abraham replies:

Babe, you do realize I am working, right?

Sarah pushes the phone away, as if toxic, scrunches her face and fights tears as she puts a small item of clothing over her face.

CUT TO

INTERIOR LUIGI'S APARTMENT- NIGHT

WEEK TWO

Sarah sits on the floor, inside Tetyana's bathroom and bathes Flori in the tub, while the two other girls, already washed and in their PJ's, run around

Luigi's apartment bathroom looks like an experiment in plumbing, with exposed PVC pipes, containers and ejector pumps all mounted on the walls rather than inside them.

Luigi stands at the door and watches Sarah play with Flori.

Without looking at him, Sarah asks, "What are all of these pipes."

"An early experiment of mine."

Sarah laughs. "Abraham is ok with all this experimenting in his building?"

"I'm not sure. He never visits," he says in Italian. Then in English he says, "Come si dici, 'What… he doesn't know… won't hurt him?"

Sarah laughs.

"It was an early prototype, for a large client," Luigi explains, "To reuse the apartamenti grey water, for the toilette."

Sarah's phone rings. She looks down at her phone, "He's calling!"

"Go, go," Luigi says. "I'll finish."

Sarah picks her phone up off the floor, stands and walks out of the room.

TIME CUT

Sarah is on the floor, in the foyer, only a few feet away from the bathroom listening to Abraham's voice over the phone.

In his English accent he continues, "...Because I am busy babe. I need to get things done. In the last few months, I've been in five countries on two continents."

"I understand and respect your mission," Sarah replies, "but our connection feels strained and so different."

Flori runs up, dressed in her nightgown. "Buonanotte!" Flori says, smiling, giving Sarah a big kiss and a hug.

"Buonanotte dolce Bambina," Sarah says and gives Flori a pat on her bottom. Then Sarah turns her attention back to the phone and in English, says, “What happened to our fantasy Abraham?”

CUT TO

EXTERIOR G.W.S. HEADQUARTERS- NIGHT

Luigi, Sarah and Caasi in Luigi’s BLUE FIAT approach the ominous brutalist concrete structure, the headquarters of the Global Water Service (G.W.S.).

CUT TO

INTERIOR G.W.S. HEADQUARTERS- NIGHT

Sarah and Caasi walk through the glass entry behind Luigi and into the double height space beyond. Caasi looks up at the broad balcony, draped in green, leafy, vegetation framing a brushed Stainless-steel water-jet cut G.W.S. Logo mounted on the face of the mezzanine.

“G.W.S.? Caasi asks.

“Global Water Service,” Sarah says in a whisper. “They track a constellation of LEO satellites that monitors water infrastructure globally.

“LEO?” Caasi asks.

“Low Earth Orbiting,” Luigi says, in English. “Remember the cinque livelli of atmosfera I taught you?”

“Yes,” she says smiling, “Tropo Strato, Meso, Termo, Esosfera. So, what level of the atmosphere are your Satellites?”

“Termo and Esos.”

“We flew through the lowest one,” Caasi says, looking at Sarah. “The Troposhere.”

Sarah smiles at Caasi. “Did we?”

Sarah, Caasi and Luigi walk, side by side, through the ground floor of G.W.S. employees, all sitting in front of monitors tracking satellite data. Then they walk up an open stair to the second level.

On the second floor, Luigi leads Sarah and Caasi toward a glass enclosed hydroponic lab. It is filled with artificial light and houses a variety of green leafy plants. Luigi stops briefly, a few feet from the lab, at the desk of a young man. He has an intricate digital model displayed on his monitor.

After studying the drawing for a few seconds, Luigi pats the young man on shoulder and says, "Bene."

Caasi looks for approval from Luigi to keep walking into the greenhouse lab. Luigi grabs her shoulders and says, "It's ok, go."

"I'm going to find my favorite," she says as she walks through the lab's open glass door.

"Basilico?" Luigi asks

"Si. Basil," Caasi says, as she walks out of site.

"What are you working on?" Sarah asks, mesmerized by a detailed construction document on the young man's screen.

"This design," Luigi says, "Is a new design to feed our new indoor vertical farms, with our water generation tech."

Sarah laughs. "The U.S. only supports Server Farms that consume our electricity, drink our fresh water and produce extraneous data ad infinitum, while Americans eat processed food, drink Coke and Pepsi and consume Ai slop."

"Do they really need that much extra for cooling?" Sarah asks, interrupting him.

"Well, those data centers do need millions of gallons a day. But this new product," the young man says, pointing to his screen, "will support our struggling farmers and lessen our ground water degradation from pesticides and fertilizers."

Sarah looks over another employee's shoulder, a female. She is speaking with Luigi. Her monitor shows indecipherable data over south east Asia when she says, in Italian, “We’ve just entered the most recent data from his Asia trip into his LLM.”

Luigi looks at Sarah to explain. “Abraham is gathering data for a novel idea.” Then to the young lady, “Bellisimo dallo a Emanuella.”

“What idea?” Sarah asks.

“I’m sure he’s told you? No?”

“No.”

“He must have?” Luigi says, “It’s all that man ever talks about.”

In almost perfect English, the young woman confirms, “This is true Signora. It is what Mr. Philipose is passionate about.”

Sarah stares at them both, with a blank expression.

“Never?” Luigi looks at her, surprised. “Abraham believes the LEO satellite debris, from the thousands of rocket launches is causing the uncontrolled counter geoengineering of the climate.”

Out of sight, Caasi calls out, “I found it! Basil. My new favorite.”

“He never mentioned satellite debris,” Sarah says.

“Luigi,” Caasi calls out, from inside the greenhouse.

Luigi walks away from Sarah and the young woman. “Andiamo Bella,” he calls out to Caasi, “To the roof.”

The young woman turns around in her green upholstered desk chair, and states, “Before that we studied how atmospheric microplastics were influencing the environment through cloud formation. It’s had a dramatic effect on precipitation.”

“He never mentioned that either?” Sarah says, confused.

Caasi and Luigi exit the greenhouse and head toward an enclosed stair.

“Of course, I want to see how your water machine feeds the farm.” Caasi says, almost out of sight. Turning around she calls out, “Coming Sarah?”

“I’ll be right behind you.”

Sarah pulls up a matching empty Morrison Hannah green desk chair and sits down, refocusing her attention on the young woman in front of her. “Can I ask you a question?”

“Of course, Signora”

“Are you in contact with Abraham?”

“Of course, many times a day.”

Sarah looks at the woman, not sure how to respond. Then asks, “His responses are…quick?”

The woman laughs, “Very. They come at all hours and are often with detailed and lengthy replies.”

Trying not to pry, Sarah asks, “And the satellites? How does he think they are adding to the global warming?”

“Satellite debris releases aluminum oxide nanoparticles into the atmosphere. It produces 400 metric tons of ozone-depleting particles a year. That’s a 700% increase in the last ten years.”

A LOUD PHONE PING interrupts their conversation. Sarah looks down at her phone’s screen which reads:

Not cool talking about me with Luigi. Yes, my family was controlling, but none of that has anything to do with me anymore.

CUT TO
LATER

INTERIOR ABRAHAM'S APARTMENT- DAY

Sarah sits in front of Luigi’s computer under a desk lamp texting Abraham:

Each text feels like a gut punch.

The room is SILENT except for one AMBULANCE SIREN that fades in and then out again. Sarah looks up to see if it wakes Cassi.

Then she continues to text three texts in a row:

Sorry if I overstepped, but contacting me only when you want a sexual exchange feels impossible.

I'd be a fool to keep feeding your ego, if there's nothing here for me.

We are suddenly a digitally incompatible couple stuck in a digital relationship.

A TEXT VIBRATION comes in:

Ugh... such a struggle Sarah. Our lives are very different.

Sarah looks at Caasi, across the room, and replies, angry:

Our lives are not different at all! We are both trying to save the world just in two very different ways.

An instant reply from Abraham reads:

This really isn't working for me.

The words, hit her stomach and send waves of pain down her arms. Tears flood Sarah's eyes.

CUT TO

INTERIOR ABRAHAM'S APARTMENT KITCHEN- NIGHT

WEEK THREE

CUE SONG: “FLUER BLANCHE” by Orsten

Sarah sits alone in the kitchen, hunched over a plate of uneaten green food. She has rings under her eyes. She picks up her phone, with its pitch-black glass, then puts it down again. Luigi and Caasi TALK in another room, while a record plays. At least three AMBULANCE SIRENS fade in and out consecutively, then Sarah picks up her phone to text Abraham;

For twelve years I have steeled myself to survive my aloneness. I built a wall, so I wouldn't hurt anymore, but like a skilled mason, you demo-ed it one brick at a time… now I'm raw and exposed and your silence hurts as much as physical abuse.

CUT TO

INTERIOR ABRAHAM'S APARTMENT STAIRWELL - NIGHT

Sarah sits alone on the steps, inside the stairwell, texting. The sound of the three children playing an Italian version of the game Hide and Seek can be heard through the door.

“Uno, due, tre, quattro….” where one counts, followed by the silence of a search and ending with squeals of delight upon being found.

Sarah’s text to Abraham reads:

```
I do have compassion for your struggle Abraham, but utter sadness to know this new pattern may never change. I'm not sure why I'm here anymore.
```

An instant response comes in. Sarah sits up, surprised.

```
Hey, hey. I'm here. Just working. It's Chaos. Pulled in too many directions. I'll reach out later ok, Gorgeous?
```

CUT TO

LATER

INTERIOR ABRAHAM'S APARTMENT- DAY

Caasi is asleep beside Sarah. Restless, Sarah gets up out of bed, picks up her phone off of the night table and walks out of the room.

CUT TO

CONTINUOUS

Sarah bumps into Luigi in the hallway.

"Are you still waiting for him to reply?" Luigi asks, in Italian.

"I thought he meant later… today?" Sarah whispers. "I just don't know why this is happening? He pulled a yearlong morphine drip and like an addict… I'm not feeling well." A muffled sound of AMBULANCE SIRENS fades in and out, in the distance. "My gut is telling me something is very, very wrong because Abraham has never been this evasive before,"

Luigi leans in, gives Sarah a hug and whispers, "Buonanotte, bella."

CUT TO

LATER

MONTAGE

CUE SONG: "PLAYGROUND LOVE", by Air

Back inside the bedroom, Caasi is still asleep while Sarah sits on the white lounge chair with her ear buds playing music. Under a floor lamp she flips through the 100+ page document that she printed out in New York.

"What did I do wrong?" she whispers, to herself.

TIME CUT

Sarah lays in bed texting. Caasi is asleep beside her.

Addicted. You called it that. Admitting how scared my reality made you feel.

TIME CUT

Sarah sits at the desk again, this time in the dark, but with her head on the desktop. Her phone screen, beside her, is black.

Still talking to myself. It's not nice to hurt me like this. Please stop.

TIME CUT

Sarah, in a fetal position, on the floor, with eyes wide open and a blank expression on her face. Her phone beside her remains dark.

I have been put on a digital starvation diet and losing fast. It was never about where you were, but how you were, with me, in this tiny little black box.

MONTAGE END

CUT TO

LATER

INTERIOR ABRAHAM'S APARTMENT- DAY

Sarah is finally back in bed, in a deep sleep, beside Caasi. The apartment is QUIET, finally with no Ambulance Sirens in the background. Out of a deep sleep, Sarah sits straight up, she clutches her stomach with a look of pain and runs out of the room.

CUT TO

Inside the bathroom, Sarah is on the toilet naked, with a small waste basket on her lap about to throw up.

CUT TO BLACK.

LATER

Sarah wakes up on the bathroom floor, beside a puddle of her own vomit, GASPING for air. She sits up feeling for a lump behind her ear, where she hit her head on the edge of the tub, she crawls into the tub on all fours. Turning the spout on, she SPITS toward the drain, while rinsing her face off in the running water.

Sarah kneels, presses the diverter for the handheld. Rinsing her body off she sees period blood mixed in the water between her legs.

"Mother Fucker!"

CUT TO

The bathroom is steamy, Sarah turns the water off and reaches for a towel. She crawls back out of the tub, the way she crawled in, on all fours. With a towel wrapped around her and her hair dripping wet, she lays in a fetal position on the small bathroom rug in front of the sink. She touches the bump behind her ear with her fingers, then closes her eyes.

TIME CUT

A loud PING from outside the bathroom rouses Sarah. Blinking, she remains on the floor, in a fetal position. When a second PING sounds. Sarah gets up, crawls toward the bedroom, desperate to connect. In transit the towel becomes undone.

CUT TO

Naked, Sarah reaches for her phone lit by her husband's glowing teepee lamp beside it. Crouched on the floor, like a feral animal, she eagerly reads the message on the screen:

`Relationships come with high expectations, Sarah. Maybe I am selfish? But I want less complexity.`

Devastated and sick, Sarah stays on the floor, naked and dumbfounded as another AMBULANCE SIREN fills the room. This time its flashing blue light makes it through the cracks of the shuttered windows.

"Are you ok?" Caasi's little voice, muffled from within a pillow, asks, while patting Sarah's wet hair with her hand.

Starting to cry, Sarah says, "I'm ok."

Caasi continues to pat Sarah's wet head as another AMBULANCE SIREN fades in, and then out. "Why do those sirens go on and on and on when everyone is asleep?"

"Because a lot of people are hurting...everywhere," Sarah says, her voice breaking.

"Don't cry," Caasi says. "I love you, even if he…" Caasi stops herself. "I just love you."

CUT TO
LATER

INTERIOR BUILDING STAIRWELL – DUSK

Dressed, but still with a head of wet hair, Sarah knocks softly on Tetyana's door. After a few minutes, Tetyana opens the door. Her blonde hair tied in a bun, wearing only a tee shirt and panties. "Is everything ok?" she asks in a hushed voice, CLICKING the hall light switch on.

As Sarah steps into the light Tetyana says, "Oh my God, you look terrible...what's that man doing to you"

"I need your help. I'm speeding toward menopause and just lost my five-month chip. Do you have a pad or a tampon?"

Tetyana laughs a little and says, "what a literary take on the insanity of the stop and start of our periods during the change." As Tetyana leads Sarah toward the bathroom she says, "Luigi and I are both worried about you."

"I'm not sure I belong here," Sarah admits, "but we can't go home anymore, that's for sure."

Tetyana opens a small cabinet, inside the bathroom, grabbing some pads and tampons. "I suppose you're officially a refugee now too?"

"Yes, the U.S. is no longer the home of the free and the brave."

"That's because," Tetyana whispers, "that free and brave woman is here now." She hands over the small pink, plastic, packages, "two bleeding refugees, in a foreign land then?"

Sarah smiles uncomfortably. "This one is bleeding profusely from the heart at the moment."

Tetyana puts her arms around Sarah and says, "We are both living proof that survival works, never without a new scar...but we do get up and keep going."

CUT TO

LATER

INTERIOR ABRAHAM'S APARTMENT- NIGHT

Sarah is alone in the apartment, inside the small library vestibule, sitting on the floor with a book open beside her, visibly upset while reading another text from Abraham:

...Oh my God babe, you cannot continue to freak out on me like this. Maybe coming was a mistake?

"Fuck!" Sarah mouths the word silently but with rage.

CUT TO

INTERIOR LUIGI'S APARTMENT- NIGHT

WEEK FOUR

Sarah sits opposite Tetyana, at her kitchen table. There is a stack of cooled crepes on the table between the two women. They both sprinkle and roll each one with sugar.

"...You cannot let a man control your emozioni."

A tear rolls down Sarah's weary face. The children LAUGH, in another room, while play acting in broken English and Italian.

"Oh cool." Liliya says, very dramatically.

"Cooo. Cooo," Flori says, mimicking her older sister.

Tetyana sprinkles sugar on a crepe and rolls it. "I have no idea what he said to you to think he's any different than he is behaving now?"

“Because there was never a text that sat, unanswered, until now?” Sarah replies. “Where did his empathy go?”

“The Abraham I know,” Tetyana laughs, “is a selfish, self-absorbed, workaholic.”

Caasi pokes her head into the kitchen in excitement and says, “Andiamo a cena adesso. That means we are going to dinner now,I’m so hungry.” Caasi turns and sees Sarah crying again. She puts her head on Sarah’s shoulder and whispers, “Please remember I love you.”

Tetyana holds out a crepe to Caasi and says, “Try this.”

Caasi eagerly bites into it. “Yummy.”

"It’s Nalystnyky,” Tetyana says.

"NAL-Y stinky?” Caasi repeats. Sarah laughs. Then Caasi runs out of the room, mouth full, but repeating the funny desert name. “Look Liliya, now I speak Ukrainian too. Nal- Y-Stinky.”

“From one widow to another, the only person that deserves your unconditional love,” Tetyana says, “... is her.”

Luigi CLAPS his hands in the other room. “Andiamo bambini!”

CUT TO

LATER

EXTERIOR ROME CENTER - NIGHT

MONTAGE

Sarah, Caasi, the girls, Luigi and Tetyana walk through the hectic city center, with CARS HONKING, AMBULANCE SIRENS coming and going and many other Romans on foot.

CUT TO

Sarah and Tetyana walk side by side as they cross over the Tiber River on the Ponte Cestio bridge over what used to be the ISOLA TIBERINA, now mostly under water. The level of water in the Tiber is still high.

Luigi and the children walk ahead of the women.

CUT TO

Tetyana holds Flori's hand as they move around the edges of Rome's historic center.

CUT TO

Liliya and Caasi are out of site, with Luigi, as Sarah trails behind all of them, looking down at her cell phone.

CUT TO

As they approach the Vittorio Emanuel monument, Caasi runs back and forth between Sarah and Luigi, excited pointing, while Tetyana carries Flori on her hip and her desert, package in the other hand.

MONTAGE END

CUT TO

EXTERIOR BASE OF THE CAPITOLINE HILL- NIGHT

At the base of the Capitoline Hill, Liliya and Caasi run ahead, up a wide expanse of steps, the Cordonata Capitalina.

In Italian, Tetyana calls out, "Don't go too far!"

Luigi takes Flori from Tetyana and puts her on his shoulders as the three adults follow the girls up the steps slowly.

"...Abraham has no time or patience to read anything other than data," Tetyana continues.

You're too hard on him," Luigi says, in Italian. "He's an engineer."

"But you're an engineer?" Sarah says. "Tetyana says you always read her writing."

"Sarah, I am fifteen years older than Abraham. I see value in many things, but most Engineers are not comfortable with the interpretative quality of art or abstraction."

"Too many excuses for this man," Tetyana says, annoyed.

"He's still too young," Luigi says, to Teyana, in Italian. "And you're too young, to be this impatient."

"If I am anything, it is impatient," Tetyana says, in English, laughing.

"You don't understand just how long it takes a man to begin to open his eyes," Luigi continues, in Italian.

"If that were true, he'd at least have become more curious about the humans around him over the past four years."

"Be nice," Luigi says, in English.

The girls disappear out of sight at the top of the stairs.

Tetyana calls out, "Don't go too far."

"I agree Sarah," Luigi says, "curious and consistent with humans does not sound like our Abraham."

Tetyana laughs.

"But we are all capable of change," Luigi says, in English. Then to Tetyana, in Italian, he says, "You are cruel."

Which makes Tetyana laugh, even louder.

Once at the top of the stairs, inside the Piazza del Campidoglio Luigi puts Flori down, then chases the two older children, who SQUEAL in delight hiding among the sculpture fragments under the colonnade.

CUT TO

LATER

INTERIOR EPSILON GmbH CORPORATE APARTMENT - NIGHT

In the Rione Monti neighborhood, overlooking the Colosseum, twelve to fifteen guests attend a dinner party hosted by a VP of

Epsilon GmbH. Each guest is connected to Global Water Service, ReitAi or Epsilon. All are of various nationalities.

The only children in attendance are, Liliya, Flori and Caasi and two German and English-speaking South Asian Indian boys, ages seven and nine. The apartment is their home.

Sarah walks toward Tetyana with two glasses of wine. She hands Tetyana one.

"You're being so quiet," Sarah asks, "Is everything ok?"

"Writing is talking," Tetyana replies, reaching for her wine, "so when I'm out, I prefer to listen."

Sarah continues across the room, toward an open French door window overlooking the Colosseum. Luigi stands with John, the Epsilon VP and host of the party. He is a German man in his early forties, and the father of the boys. Beside him is an older man, with a jovial facial expression, an Israeli expat, with a heavy accent, speaking English.

"...People hate truth tellers," says the expat, "that's why I'm hated."

"…You are not hated, Sam." John says, in German.

"On the contrary," Sam says, in English, while looking at Sarah, as she approaches, "I haven't spoken to everyone here yet." To Sarah, Sam says, "I love to be hated, and I hate to be loved."

Sarah looks at Sam confused.

"Stop flirting," John says, in English, in his German accent. "Sarah, this is our tenant, Sam."

Sam reaches out to shake Sarah's hand.

"Abraham's friend from New York," Luigi adds, in English.

"Ahhh, so Abraham does have friends?" Sam says, laughing. "And from New York, no less. A place that is longer recognizable I hear."

"The entire country actually," Sarah corrects, looking out of the open window at a commotion next to the Colosseum. She squints a bit. “Are people living inside the Colosseum?”

“The Government doesn’t know how to protect all of the climate refugees,” John says. “There are just too many of them.”

Together the four adults watch the swarm of humans a few hundred yards away and their ant like movements surrounding at least twenty of the Epsilon Atmospheric Water Generating machines.

“My God.” Sarah looks over at Luigi. “There are thousands of them.”

In Italian, Luigi says, “I’m sure you’d prefer that to the catacombs?

“Catacombs?” Sarah asks, looking at all three men.

“He’s not lying,” John says, “they run for 150 kilometers, at 20 meters underground and stay very cool, even during the hottest daytime temperatures.”

CUT TO

The three older children, Caasi, Liliya and one of the boys, run out of the kitchen, into the living room, toward Luigi and Sarah. As other two run off, Caasi stops and tugs on Sarah’s arm. “I’m so hungry.”

Sarah bends toward her. “Yes, its late, let’s go see what they have.”

Sarah holds Caasi’s hand and walks away from the men toward a buffet being prepped in a large room near the kitchen.

Veda, the hostess, an English woman, originally from Calicut, India and two servers put the final platters of food out, each more colorful than the next. Caasi looks around at the food, dismayed.

“So many fresh vegetables,” Sarah says, to Veda, “Unheard of in the States anymore”

“Luigi’s vertical farms have changed so much.” Then she turns to Caasi, and says, “I think you’ll like this dish.” Veda points to a rice dish.

Caasi looks at it with curiosity.

"It's Biryani?" Veda says, hoping for some recognition.

Caasi looks up at Sarah.

"She's never had Indian food." Sarah says.

"Never? Una buona giornata per iniziare." Translating her own words Veda adds, "A good day for a young Indian to start then." She puts a small amount on a plate and hands it to Caasi. "It might be a bit spicy, but it's my son's favorite."

Just then her youngest son, runs up and wraps his tiny arms around his mother's waist. Veda, smiles, gives him a big hug and walks with him back into the kitchen.

Caasi takes a bite and chews the rice dish tentatively, but makes an instant face. "It's so spicy." Caasi whispers, trying not to be ungrateful.

"It's hard when everything is new," Sarah takes the small dish from her. "You don't have to eat the rest. I all the deserts at the other end of the table. Go grab one before dinner."

CUT TO
LATER

Two guests, an Italian female, in her late thirties and Italian man chat in a doorway at the edge of the dining room.

"...The incidence of miscarriage," says the woman, "was increased by 25% during the COVID-19 pandemic."

"I had no idea." The man says, in Italian.

CUT TO

Seated at the long dining table, John, the host and an Epsilon employee, argue with a balding, opinionated, American man in his forties. The man speaks with a heavy southern drawl. Luigi approaches with two plates of food for Liliya and one of the boys already seated at the empty dining room table.

"...vertical farming, of course." The Epsilon employee says, in English.

"But undermining farmers?" the American, says. "That just wouldn't fly in the US."

"Not quite undermining," John says, trying to be diplomatic. "Just introducing a more sustainable model. One that uses ninety percent less water and no pesticides."

CUT TO

Flori sits on Tetyana's lap leaning over a plate on a side table in the living room. The three-year-old is hyper focused on trying to get one last green pea onto her spoon and is chasing it around the plate as Tetyana and an older woman look on.

"The 'Y chromosome', that create male offspring, may be faster…," Tetyana says, to an older woman, "but its female 'X' counterpart, lives longer and is therefore much more persistent."

The older woman laughs.

"To answer your earlier question," Tetyana says, "yes, I do write at all hours. Because words are fleeting, you must catch them like butterflies."

CUT TO

At the buffet, Sarah and Luigi talk, while filling their plates. The American man, with the Southern drawl, stands behind Sarah, smelling most food before putting it on his plate.

"…there are patterns," Sarah continues, "even in chaotic systems."

"I don't think we are talking about tomatoes anymore." Luigi says, in Italian.

"Sorry, it's the teacher in me," Sarah replies, in English.

"Teacher?" The fussy man perks up. "Any place I'd know?"

Sarah looks at him, suddenly wary of the American accent. "No. Only here in Roma."

Sarah walks away, Luigi follows close behind.

TIME CUT

A group of guests, seated at the dining table, participate in an intense discussion. Luigi sits at the last open seat at the table. Sarah eats while standing, listening.

"…The only reason our newest AGI, is such a success is because it finally falls within the EU guidelines," John says.

"You mean guardrails," interrupts a young man with a ReitAi Tee shirt on.

"Yes, Helmut, it is safe because of its need for a human to feed it singular datasets for singular problems."

"But it's still an Artificial General Intelligence machine?" another guest asks.

"Indeed," John says.

"So, it does have independent thought then?" the same guest asks. "And capable of independent action?"

"Yes, but the safety protocols, agreed upon at the G7 in 2027, have finally been met." John tells the group.

"Didn't the United States host the G7, that year?" another guest asks.

John laughs. "That event is why most Big Tech moved their operations, data centers and headquarters to Europe. The US stood in defiance to the EU Ai Act of 2024, that most of us eventually agreed upon."

"But the chaos of a tangled hose or electric cord," Sarah says, "will never be solved by any Ai or AGI. Only a human brain can untangle that kind of complexity."

"Comparing a brain to a computer is an insult to our brain," says one female guest, with an English accent.

"She didn't say normal brain," Sam, chimes in, "Perhaps Ai is like a narcissist's brain?"

John laughs. "Of course, you would find a way to turn this into a conversation about narcissism."

"Normal brains are cognitive misers," Sam replies. "They use the minimum amount of energy necessary to solve a problem, while a narcissist would over invest in an attempt to appear perfect."

The entire table laughs. Caasi walks up to Sarah, carrying Flori on her hip.

"You're calling us cognitive misers?" the same woman asks. Then switching into Italian, she adds, "This room, holds some of the smartest minds working to save us from ourselves? It's an insult to be so reductive."

After a pause where the group laughs and Sam's dinner neighbor, translates what was just said, Sam says, "When Ai lies, like you do invariably, it is human."

There is an uncomfortable silence in the group.

"Let's settle down Sam," John says. "No one wants the evening to end on a sour note."

Caasi listens with curiosity, Flori sucks her thumb and leans onto Caasi's shoulder, then Caasi leans into Sarah's abdomen.

A male guest, standing beside Sarah, looks at Caasi cuddle, and asks, "How old is he?"

Caasi's dark short hair is tucked behind her ears, and her outfit of shorts and sneakers, belie her true gender.

"Twelve," Sarah whispers.

"I wish mine were that curious and loving."

"Intellect can to be nurtured." Sarah says.

"So does love, apparently. Brava," the man says.

Caasi walks away when one of the younger boys, runs by laughing and pokes her in the rear.

In a statement directed at John, the American with the southern drawl, says, "At least a narcissist's brain doesn't need extreme cold to function, like ReitAi's Quantum processors do." Then he adds, " Epsilon's backing of their investment in Antarctica, was genius, but will it stay cold enough? With this global heat?"

Put off by the unintelligent remark, John's states "We aren't in Antarctica because of the cold. We're there for the environmental silence needed to prevent quantum decoupling."

CUT TO

EXTERIOR ANTACTICA MEGACITY- DAY

A massive multibuilding mini city sits on a plateau, on a sheet of white ice, in the middle of a snow-covered landscape with a bright blue cloudless sky overhead. The site is SILENT with no humans or movement in sight but it is surrounded by ten yards of barbed wire as far as the eye can see. A small placard reads:

GLOBAL QUANTUM PROCESSORS

Managed by ReitAi GmbH

Developed by EPSILON GmbH

In agreement with

The 1959 Antarctic Treaty System

This property is being used for scientific purposes only

CUT TO

LATER

INTERIOR EPSILON GmbH CORPORATE APARTMENT - NIGHT

Sarah walks past the dining room, as her PHONE RINGS. Luigi looks at her, from the table, as she answers it and walks away.

CUT TO

CONTINUOUS

INTERIOR CORPORATE APARTMENT, BEDROOM- NIGHT

Sarah picks up the call.

"...I was just fantasizing with you on my mind," Abraham's voice says.

"What happened this week?" Sarah asks softly, as she continues to walk away from the party "And last, and the week before that?"

Sarah stops at an unlit, unoccupied bedroom, goes in and sits down on the edge of a bed. Far from the party's chatter, she listens to Abraham's unfamiliar voice.

"I'm here now, babe."

"This is so hard," Sarah admits.

"Tell me why?"

"No time to connect."

"You know I've been working very hard. I barely have any time for anything."

"But you want me when you want, and don't when you don't want? It's not fair. Your behavior, it's so… off."

"Nothing's changed, I still want you. I'm here, thinking of touching you all over."

With her foot, Sarah pushes the door closed, and lays back onto the bed, simultaneously.

"Let's just be online together now," he says, "I love that we connect when our stars are aligned. Hearing your voice drives me wild."

"I'm so thirsty for your kindness," Sarah says, "Please be gentle and kind."

TIME CUT

Sarah rejoins the party after her call. Many of the guests have resettled. The group sitting around the dining table is much smaller. Sarah smiles at Luigi and she moves in to sit beside him at the table.

"...I said this in 1995!" says Sam, defensively.

"You certainly don't exhibit an ounce of humility," says an English woman.

"Insufferable and disagreeable more like it," says the American, with the southern drawl.

"The pot calling the kettle black," someone else mumbles, loud enough for all to hear.

Everyone laughs.

"Turing was wrong," Sam says, doubling down.

"In what way?" asks the English woman.

"In believing that Ai could pass as human."

"They're not designed to be human," Helmut, the man with the ReitAi Tee shirt on, says. "Just expert at solving human problems."

"Ai and even your newest AGI are no experts," Sam says, "They are just like a narcissist, they steal ideas and misattribute those ideas to themselves."

"Sounds like an admission, coming from a self-described narcissist?" says another woman, with a French accent.

Everyone laughs again, even Sam.

"We may be in the midst of the Fourth industrial revolution,"

Sam says, "but a Narcissist, just like Artificial Intelligence is a pond pretending to be an ocean. Neither has any idea what they're talking about."

Caasi comes up behind Sarah and whispers into her ear. "Where were you? I was looking all over for you? That man was asking questions about New York."

"Which man?"

"That man," Caasi whispers. "The one across the table," pointing to the American with the southern twang across from Sarah.

"Did you tell him anything?"

"No, I just acted like I didn't know what he was talking about. And ran away with the boys."

"Get Liliya and Flori." Sarah leans toward Luigi, touches his arm and in Italian says, "I think we should leave."

CUT TO
LATER

INTERIOR ABRAHAM'S APARTMENT- NIGHT

Sarah is in the bathroom, Caasi is babbling about the party through the cracked door,

"... it was fun because I miss all the boys at school. They play different."

Sarah walks out of the bathroom, in a nightgown,

"Go get into bed, I'll be there in a minute." Sarah tucks her head around the corner into the kitchen, where Luigi sits, texting.

"John just replied," Luigi says, in Italian, "He was a diplomat, from the U.S.A."

CUT TO

Sarah walks into the bedroom. Caasi is in bed, but starts chatting once Sarah enters.

"Did you hear me?" Caasi asks, "I was trying to say that play between girls is very different than between boys."

"That is true." Sarah says, while sitting down at the desk to put cream on her face, replying to Caasi and simultaneously sends a text to Abraham which reads:

`Something strange happened tonight. We may not be safe here anymore.`

"...but both types of play have value." Caasi says. "Don't they?"

TIME CUT

Sarah is fast asleep in bed, Caasi is missing. Through the open bedroom door Caasi and Luigi can be heard CHATTING in the kitchen in English and Italian. Sarah's phone VIBRATES, unexpectedly, on the small stone night table. The time on the screen shows, 8pm. The Phone VIBRATES again, which rouses her out of a deep sleep. She reaches to answer it.

Without a formal greeting, Abraham says, "What do you mean not safe?"

"Where are you?" Sarah asks, still waking up. "Please come soon."

"I'm in Sicily."

"In Italy?" Sarah says, sitting up.

"I thought I told you?"

"No... you haven't been clear in weeks." Sarah pauses, dumbfounded. "You said you were busy, bouncing around, but not with what or where."

His reply is instant. "I really don't like being micromanaged."

"Seriously?" Sarah waits for a reply, that doesn't come. "What's happening?" She looks at her phone, sees he is still on the line, then puts it up to ear again. “Abraham?”

“I’m afraid of what will happen,” he says.

“When?”

“Post contact.”

Sarah listens, but doesn’t reply.

“My emotions will tumble,” Abraham admits, “I’ll want to be with you all the time and that’s just not scalable.”

Sarah remains silent.

“You wanted me to be honest,” he adds, “This is me being honest.”

Sarah takes a deep breath, with a tear in her eye she says, “I have to go.” And without waiting for a reply, she hangs up the phone turns the ringer off and tucks it under her pillow.

CUT TO

CONTINUOUS

INTERIOR ABRAHAM’S APARTMENT, KITCHEN – NIGHT

Sarah walks into the kitchen. Luigi and Caasi are playing cards over breakfast. Luigi looks up at Sarah.

“He’s in Italy,” Sarah says, in Italian.

To Caasi, Luigi says, “Go find the record we were talking about.”

Caasi leaves the kitchen.

“You knew?” Sarah continues, in Italian.

“I assumed you did too?”

Sarah stands in the doorway with tears running down her face.

Luigi gets up from the table. “Oh, Sarah. Abraham is complex. His ability to communicate outside of work and in relationships is a struggle.”

Sarah sits down at the table and puts her elbows onto the table and her head into her hands,

"I'm sure if he's in the country he'll be here soon," Luigi says, in Italian.

"It's been four weeks," Sarah says in English, without looking up, and then adds, "…of torture, Luigi. Pure torture."

Luigi squeezes her shoulders and says, "I'll be right back."

After a few minutes Luigi walks past the kitchen with Caasi and then out of the apartment. The door SLAMS SHUT.

SMASH CUT TO
LATER

INTERIOR ABRAHAM'S APARTMENT- NIGHT

The apartment door CLICKS open, then SLAMS SHUT. Sarah startles, from a fetal position on her bed. After a few seconds Luigi enters her bedroom, alone.

"Let's go out," Luigi says, in Italian "I'd like to show you something."

Sarah turns around on the bed and sits up. "Where's Caasi?"

"She's playing with the girls for the evening."

"It's ok to leave her there?"

"Of course."

CUT TO
LATER

INTERIOR LUIGI'S CAR - NIGHT

CUE SONG: "MOON RIVER" by Henry Mancini orchestral version

The radio plays softly, as Sarah and Luigi drive out of Rome.

"At its height, eleven aqueducts, fed Rome, from as far away as ninety-two kilometers," Luigi says.

Half listening, Sarah looks out of the window watching the lit city landscape transform as they enter the dark country side.

"...Romans built over two hundred aqueducts across Europe, the Middle East and North Africa..."

Still half listening, she turns to watch the orchestrated movement of Luigi's right hand on the stick shift and his left foot on the clutch as he continues.

"...the Aqueducts were built over a period of five hundred years, and some still feed Rome its fresh water."

SMASH CUT TO
FLASHBACK

INTERIOR CAR, SOMEWHERE IN ITALY- NIGHT

15 YEARS EARLIER

With his right hand on the stick shift Nicola, Sarah's husband speaks feverishly in both Italian and English.

"Such a perfect town...," Nicola says, in English, then in Italian he adds, "springs, rivers and hydropower. When we have our first child, we'll buy something here."

Snow covered mountains can be seen in the distance, glowing from the moonlight. Light rain hits the windshield as they drive through the dark cobble stone streets. Passing a red horizontal street sign, with its white outline and white letters, that read:

DOMODOSSOLA

citta per la pace

(translated 'a city for peace')

FLASHBACK END
TIME CUT

EXTERIOR CAR, SATURNIA- NIGHT

Luigi's car drives down a dark unpaved dirt road through the moonlit countryside. They arrive at a gate.

CUT TO

CONTINUOUS

INTERIOR CAR, SATURNIA- NIGHT

Through the windshield a sign in white letters reads:

SATURNIA

Parco Termale

Luigi stops the car, but not the car's engine.

"What is this? Sarah asks.

"A hot spring," Luigi says, as he opens the car door and gets out.

CUT TO

CONTINUOUS

EXTERIOR GATED PROPERTY, SATURNIA- NIGHT

Luigi walks over to the locked gate and unlocks its padlock with a key.

Sarah opens her car door, stands up, half in and half out of the car and calls out, "You own this?"

The sound of RUSHING WATER is loud, but its source cannot be seen.

"Non, Bella, but my team and I monitor it." he says, focused on the lock.

CUT TO

Luigi parks the car inside the gate, then locks it again.

CUT TO

With a small bag over his shoulder, Luigi walks beside Sarah across an open moonlit field toward the steam that is rising along its horizon.

CUT TO

CONTINUOUS

CUE SONG: THE PARKING LOT by Andersen. Paak

Luigi reaches out to hold Sarah's hand. Sarah looks at him, then with a smile holds his hand in return. Standing on the horizon, they look down and see irregular bodies of water stepping down along the hillside, like rice paddy terraces, each irregular in shape, glowing in an unearthly milky blue.

"A gift for you," Luigi says, in Italian. "For your forty-fifth birthday, the one that Abraham forgot."

A tear rolls down Sarah's cheek, which she wipes away with the back of her free hand. "I've been so emotional this month. I have no idea what's happening to me."

TIME CUT

On the other side of the hill, close to the water, Luigi lays out a blue checked blanket. Able to see the water's source, and a small stone structure beside it. The terraces seem like one living, breathing object.

"We are standing on a fault line," Luigi says, in Italian, "an impending tectonic shift."

"Poetic," Sarah says, "And somehow appropriate."

Luigi laughs, and in English he says, "I come here often when I need to breathe."

"Not the sulfur, I hope," Sarah laughs.

He laughs. "Romans believe the water is healing."

TIME CUT

Luigi and Sarah sit beside each other, laughing and share wine, directly out of the bottle.

"…All hot springs are beautiful," Luigi says. "Each one surprises me in unique and magical ways," then after a pause he adds, "not unlike women."

Sarah looks directly at him for a few seconds. "This will be a forever memory…Thank you."

"Memories," he says, "are the only gifts worth giving."

TIME CUT

Sarah jumps to her feet, excited, playful and tipsy. She immediately starts to remove her clothing. "We're going in right? It's so hot out tonight."

Luigi, sits up. "No, no the springs have become too acidic," he says, in Italian.

Sarah stands frozen, half naked, looking at Luigi, with the moon behind her.

"Please don't stop...," Luigi whispers, as he also stands up.

Sarah continues to stare at him, while slowly removing another piece of clothing.

"The raw heat," Luigi rambles on, in Italian. "And chemical reactions that our planet is made of is exposed in these bodies." He reaches out to hold one of Sarah's hands and spins her around, looking at her naked body in the moonlight. "I like finally seeing all of you." he says, in English.

TIME CUT

Both naked, locked in an embrace on the blanket, Luigi and Sarah kiss deeply as their limbs are intertwined and move gently in unison.

CUT TO
LATER

INTERIOR ABRAHAM'S APARTMENT - NIGHT

The dawn light filters through the shuddered apartment windows. Luigi and Sarah roll into the empty and unlit apartment kissing. Sarah is laughing, still tipsy, and pulls away, squeezing her legs together, saying, "I need to pee."

"...Quick. Quick, beautiful." Luigi replies.

Sarah leaves him in the small unlit bedroom beside the kitchen, where he flops onto the single bed.

CUT TO

Sarah returns to Luigi and stands in the doorway, this time with a straight face. "This was a mistake," she says, in English.

"There are no errori." Luigi says, in half English as he sits up on his elbows.

"I haven't been with anyone in ten years," Sarah admits.

Just then Luigi's phone PINGS. His face glows as he looks at the screen, but CLICKS the screen off. Then his phone PINGS another two times, which he ignores. "I don't know what it is about you?" he says, in Italian. "It's too much to resist."

"Aren't you going to answer?"

"No."

"Why do men not reply? Who are you ignoring?"

Luigi points to the ceiling.

"You're obviously involved with her."

Luigi sits up straight and in Italian asks, "Am I?"

"Yes, you are, in one way or another. Respect her and reply. Tonight was a momentary lapse. I was so hungry for physical attention. I'm so sorry." Sarah says, as she begins to walk away.

Luigi jumps up and grabs her arm and pulls her toward him. "I am not sorry," he says, in English.

Sarah looks at him for a few seconds, but pulls away and starts to walk away.

"What happened in the last five minutes?" Luigi asks, in Italian.

Sarah turns, looking down at the phone in her hand. "He's coming tomorrow." Then she turns to go into her room.

CUT TO

10 HOURS LATER

INTERIOR ABRAHAM'S APARTMENT- NIGHT

CUE SONG: "INTO DUST" by Mazzy Star

After a fitful sleep, Sarah stays in the apartment waiting for hours for Abraham's arrival.

MONTAGE

Sarah doesn't respond to Caasi, who is talking to her in the bedroom. Sarah just folds clothes, while Caasi babbles.

TIME CUT

Sarah walks from the bedroom to the kitchen, looking at the blank screen on her phone.

TIME CUT

Eating a meal, with Luigi and Caasi, in the kitchen Sarah doesn't eat much, she just stares up at the ceiling.

SMASH CUT

Sarah looks down at her phone, then up again and is sitting alone at the kitchen table.

Caasi's voice calls out, "Bye Sarah."

The apartment DOOR SLAMS SHUT.

TIME CUT

Sarah strips the bedsheets.

TIME CUT

Sarah scrolls on her phone, while laying on her back on the unmade mattress.

TIME CUT

Sarah puts the bed sheets back on the bed.

TIME CUT

Sarah makes a bed on the floor of the small library nook off of the entry, while looking at her phone, propped up on a book shelf.

TIME CUT

Sarah spins around, looking at herself, in the wardrobe mirror. Still spinning, Sarah is wearing one, two and three different outfits.

TIME CUT

Sitting at the bedroom desk, wearing yet another outfit, Tetyana's original, much tighter green dress Sarah puts on a red lipstick. It matches the red of her wedge shoes.

The DOOR SLAMS SHUT.

Luigi and Caasi, walk into the apartment laughing.

Luigi calls out, in Italian, "We'll see you later."

Caasi follows and says, "Bye Sarah."

The DOOR SLAMS SHUT.

Sarah looks at herself in the mirror again, then wipes the lipstick off with a tissue.

TIME CUT

Sarah walks toward the open window, in the small library nook and looks down at the bustling, amber lit street.

SMASH CUT

Struggling to watch the people walking by, Sarah pulls herself back in, but is now in at the kitchen window.

TIME CUT

Still in the kitchen, Sarah is washing a plate in the sink, beside the shuttered window. The front door opens, then it SLAMS SHUT.

Sarah turns the water off, listens to the movement in the foyer.

"Caasi?" she calls out. Then she grabs a dish towel for her hands.

"It's just me," Abraham says, from the foyer.

Sarah tries to dry her hands quickly, but turns to find Abraham already in the kitchen doorway, well-dressed and striking.

Sarah freezes, dish towel in hand.

"No words left?" Abraham asks.

"Too many actually."

Abraham walks toward her. She puts the dish towel down on the table. Abraham politely greets her by kissing her on each cheek, but full of emotion, Sarah wraps her arms around him, in a silent embrace.

Abraham whispers, awkwardly, "You're taller than I imagined."

MONTAGE END

CUT TO
LATER

CUE SONG: "ET MOI" by NYM

At one end of the dark corridor, the bedroom door is closed. Sarah is LAUGHING, as the RECORD PLAYER PLAYS. The front door SLAMS SHUT.

Luigi walks into the dark hallway with Caasi trailing behind him. He knocks on the bedroom door.

"Entrez," Abraham says, in French, being playful

Luigi opens the door.

In the shuttered yet, brightly lit room, Abraham sits in his white Eames lounge chair, while Sarah sits beside him on the green Persian rug.

"Benvenuto a casa!" Luigi says, happy to see his friend.

Abraham smiles, gets up to greet Luigi, shakes his hand, then they embrace, laughing.

Caasi stays behind, standing inside the door frame.

"What's it been?" Luigi says, in Italian.

"Four months, this time," Abraham says, in English.

Sarah gets up and walks toward Caasi. “Come say hello,” she whispers, holding out her hand.

Caasi shakes her head no and stays in the doorway with her arms folded across her chest.

Luigi turns toward Caasi and in English, says, “This is Abraham, our friend.”

Caasi stays in the doorway, frozen.

Luigi walks toward Caasi, “Come… get your book. And we’ll go back upstairs.” Then to Sarah, he says, “Did you know she’s reading *Breakfast of Champions*?”

Surprised, Sarah says, “No. I did not.”

Caasi walks past Abraham and bends down to pick the book up from under the bed. Then she walks past the two adults and toward the open door. Luigi puts his hands on her shoulders to slow her down, then follows her out of the room.

“We’ll catch up domani!” Luigi says, closing the bedroom door behind him.

Abraham sits down on the bed and Sarah positions herself, provocatively, in front of him between his legs. Caasi barges back into the bedroom, unexpectedly, and walks toward them to grab her doll. Without stopping, she says, “Vonnegut says, fucking makes babies.”

“Caasi,” Sarah says, “be polite.” But the DOOR SLAMS SHUT cutting Sarah’s words off.

With a large smile, Abraham puts his hands on her hips, pulling her toward him and says, “She’s not wrong.”

Sarah laughs, out loud. Grabbing his face with both hands, she bends down and says, “I am far too old for that to happen.” Then she kisses him deeply.

He pulls her down beside him, onto the bed and whispers, “My God, your spontaneity and sensuality are better than any digital avatar. You aren’t anything like I imagined. Such a firecracker.”

CUT TO

LATER

Sarah and Abraham lay naked under the sheets. The apartment is SILENT but it's dark outside again. Sarah is on her back, still asleep. Abraham is propped up on one elbow looking down at her, while he caresses her upper arm, with his fingertips. Sarah, opens up her eyes and smiles.

"Did you get these here?" Abraham asks, circling her tattoo.

Sarah laughs. "No. I've had these since I was a teenager."

"Really?"

"Yes. Why?"

Abraham turns around to pick up his phone. "They aren't in any of the pictures?"

Sarah sits up, covering her breasts. "What pictures?"

Handing her his phone, he says, "The ones you sent me."

Sarah scrolls through a collection of faceless erotic female nudes.

"Who is this?" she asks.

TIME CUT

Abraham sits, fully dressed, silent and uncomfortable, at the desk, while Sarah, also dressed, paces barefoot inside the bedroom.

"...you said we were going to create a new world together. That I held all the answers...that I was what you had been looking for...that my work was important and that together we could make a difference." Sarah walks toward her wardrobe, opens it, rummages inside. "You told me you loved me, not once, every day for over a year. You even said I held the key to your survival, which, I'll admit, felt dramatic, but charming." Sarah pulls out the small back pack. " You admired my work, were always thirsty to know more about it. How it worked. What my lessons taught."

“Was I?”

Increasingly upset, Sarah says, “Yes, there was never a day that went by where you weren’t there, asking questions, showing interest.”

“Not one?”

Sarah walks toward Abraham, with their yearlong text thread, the document she printed out on the final evening in her apartment, and says, “No, not one day!” then drops the bound digital conversation, onto the desk in front of him.

They both fall silent, as she watches him flip through the pages.

“Morning, noon and night,” she adds.

After another long pause, while Abraham studies the document, she adds, “You put me on a pedestal, hard not to fall from, but here you are, a man whose reality, it turns out is nothing like the fantasy you’re holding in your hands, yet, I was willing to throw caution to the wind, throw my life away and work to see who you really were outside the walled safety of that fucking device.”

“Sarah,” Abraham says, not looking up from the document. “I didn’t write any of this.”

“Yes... you did!”

“I don’t know what this is? You and I only had a loose sexting relationship for about six months. One I admittedly liked better than most but… that’s it.”

“But we’ve been connected for a year?” Confused, Sarah sits down on the edge of the bed and whispers, “But why did you keep asking me to come to Italy then? To come here?”

Abraham turns to look at her, “I didn’t.”

Sarah starts to cry. “What’s happening?”

“Four weeks ago, you told me you were coming, needed help

getting out of New York, help with passports and needed a safe place to stay."

"Is any of this real?" Sarah whispers to herself again.

Abraham gets up suddenly and walks toward the bedroom door. Sarah stands up to follow him, raising her voice, "Is it? Is any of this real?"

Without responding, Abraham grabs a set of keys, out of a kitchen drawer, then while still in motion, says, "I don't know, Sarah. I think we've been played."

"By whom?" Sarah says, following him out of the kitchen.

"I don't know," Abraham says, as he walks past her toward the apartment door. "But I'm going to find out."

CUT TO
CONTINUOUS

INTERIOR STAIRWELL – DAY

Luigi climbs the stairwell steps with a fresh loaf of bread in his hand. He hears an apartment door SLAM SHUT and looks up just as Abraham rushes past him, then out of the building.

The building door closes slowly followed by a heavy THUD and a CLICK.

CUT TO
CONTINUOUS

INTERIOR ABRAHAM'S APARTMENT- DAY

Luigi walks through the front door of Abraham's apartment, easing the door shut. Caasi watches him, sulking on her make shift bed on the floor, a few feet away.

"Is everything ok?" Luigi asks, in English.

"No."

"Did you apologize?"

"I came in and heard arguing," Caasi says, "He left and she is crying again, so…no. I did not."

Luigi walks away from Caasi and toward the bedroom door, which is ajar. He knocks and pushes it open without waiting for a reply. Sarah is face down on her pillow and instantly looks up, wiping her tears with her hand. "Where did he go?"

"The only place he ever goes," Luigi says, in Italian.

She looks at him confused.

"Work," he adds, in English.

"I'm such a fool," Sarah whispers. "I'll never understand."

Luigi walks toward her and sits on the edge of the bed.

"What happened?" he asks.

"I don't know…" Sarah cries out, then buries her head into a pillow.

Luigi puts his hand on her back. "Some men don't do empathy well," he says, in Italian. Then bending down to look into her eyes, says, "It took me a long time."

Sarah smiles at him through her tears.

In English he continues, "Women need kindness… how do you say…empathia and consistency."

"Yes, empathy, especially after we expose ourselves physically," Sarah says.

"Dategli Solo Tempo." Luigi says.

"How much more time do I give him?"

"He still has a lot of…," Luigi pauses, to find the right English word, "how do you say, Maturando?

"Maturing? Sarah says, "Even at thirty-six?"

"Love is not a faucet; you can't turn it off." Luigi says, in

English.

"Or on…apparently," Sarah adds.

Luigi put both his hands on her wet cheeks. "You are the first woman he has ever invited into his home. You are very important to him."

Exhausted, Sarah wipes her eyes.

"Let's get you and Caasi out of here for a while."

CUT TO

LATER

The three children sit on Caasi's bed and put their shoes on. Luigi is packing a picnic in the kitchen, while a VACUUM is running upstairs.

Sarah walks into the kitchen, her eyes still swollen from tears. "I thought she was writing?"

Luigi laughs. "Cleaning is part of Tetyana's process. She says it quiets her mind." Luigi steps past Sarah, and pokes his head around out of the kitchen door. "Andiamo bambini."

CUT TO

EXTERIOR PARCO DEGLI ACQUEDOTTI- NIGHT

Caasi, Sarah, Luigi and Flori, dwarfed by an ancient aqueduct, twenty-eight meters tall that stands behind them, finish eating a picnic. Lit by moonlight many other roman families, are scattered sitting on their own picnic blankets. Liliya plays with other children, nearby.

"...but the book is a history book," Caasi says, adamant.

Sarah and Luigi look at each other.

"It's about America," Caasi adds. "It says that sea Pirates were white and American Indians were copper colored."

Sarah smiles. "That may be but, *Breakfast of Champions* is not age- appropriate reading."

"But I have free will," Caasi says, "just like Dwayne Hoover."

Luigi laughs, under his breath. "Indeed, we all do, but perhaps Sarah is right and Kurt Vonnegut's work is not appropriate?"

"The book…," Caasi says, but stops suddenly when Flori interrupts.

"Water, please," the baby annunciates in clear English.

They all smile at her use of the new language and Luigi hands Flori her sippy water cup

"In the book," Caasi continues, "it says that humanity deserved to die horribly because it was so mean to our sweet planet."

"Mr. Vonnegut isn't wrong," Luigi says.

"On Planet Lingo," Caasi adds, "three creatures become extinct because they destroyed their planet's resources, including its atmosphere."

Sarah sighs deeply just when Liliya runs up to Caasi. "Come, Caasi! Acciapparella,"

Liliya pulls Caasi toward the other children, playing a game of tag.

"Uno momento," Caasi says, shaking Liliya's hand away. Turning back to the adults, she says, "I'm about to be thirteen. I want to feel like I can read the things I want…Ok?" Then she gets up, holds her hand out to Flori, who joins her eagerly, sipping cup in hand. They walk away, toward other children.

"She is about to become a handful," Luigi says, when Cassi is out of ear shot. "But in a good way."

CUT TO

LATER

INTERIOR ABRAHAM'S APARTMENT – NIGHT

Luigi and Sarah sit at the kitchen table. The children can be heard SINGING, 'Row, Row, Row Your Boat', in English upstairs, when the apartment door opens, then SLAMS SHUT.

Abraham walks into the kitchen and unceremoniously, without words, drops two newly printed volumes of bound paper onto the table, a smaller and one larger one.

Sarah flips through the larger manuscript, Luigi the smaller.

"This is our text thread," Sarah says.

"Yours, yes," Abraham says, "but not mine." Abraham points to the one Luigi is looking through. "That one is mine."

Sarah looks over at the smaller volume. And then back up at Abraham's face.

"Our connection was not with each other." Abraham says. Confused, Sarah does not say anything. "It was an Ai manipulation." Abraham adds.

"Whose Ai?" Luigi asks, in Italian.

"Epsilon and ReitAi's newest AGI machine," Abraham says, with a straight face.

"Ours? Are you sure?"

CUT TO

INTERIOR ABRAHAM'S APARTMENT- NIGHT

Luigi, Abraham and Sarah sit in the large back bedroom. Caasi and the children can be heard RUNNING up and down the hallway.

"They finally did it," Abraham says, stunned, "Epsilon and ReitAi finally made their AGI autonomous…it obviously tried to solve the problem I asked it to a year ago, but without any data."

"What do our private conversations have to do with that?" Sarah asks.

Still studying both documents, Luigi says, "It obviously wanted you to connect?"

"Why?" Sarah asks.

"Without data, it must have looked for a human that could help?" Abraham says. "Is it you or Caasi?"

"Fascinating," Luigi says, while studying the documents side by side. "It gave Sarah attention, love and kindness and Abraham, a no strings attached, sexual connection. After a short pause Luigi asks, "Did you ask it any questions today?"

"Sure, why it used deception! Instead of a direct request."

"And?" Sarah says, annoyed.

"It said that humans rarely respond to direct requests from strangers."

CUT TO

In the kitchen, Luigi and Abraham sit side by side, across from Sarah eating a meal together. The sound of children playing upstairs, has quieted.

"…Caasi came into their care as a healthy baby, who scored high in intellect and compassion and became central to their Digital Empathetic Twin Program, the government's race toward a viable AGI."

"You never mentioned this?" Luigi says, in Italian.

"That's because the unregulated data collection of a child's SSL, their self-supervised learning and their innate psychological and emotional milestones, in order to feed an Ai, wasn't relevant. We were here now."

"Is that what they were doing with Caasi?" Luigi asks.

"Yes, and millions of other children too. Just like in China. Education became relentless testing, not much else," Sarah says, "But they owned Caasi since birth so, she was more valuable than most. Plus at twelve she is at the height of natural empathy, intuition and intelligence, not yet corrupted by societies need to manipulate, impersonate, lie, future fake or heart break."

"I thought you said she's only been with you for a year?" Abraham asks.

"Yes, to engage her intellectually and train her my lessons, but she's been under the government's microscope since birth."

"The Eliza effect, on steroids?" Luigi says, in Italian.

"More like a skinner box," Abraham says.

"The US wants their AGI not to be gullible like a toddler, but not to hallucinate; lie, cheat and steal, like most older children and adults. So, they have been studying the moment innocence is lost.

"Joseph Weizenbaum was not wrong to call Ai 'an index of the insanity of our world'," Abraham says.

"If the EU has guardrails in place, why would Epsilon's AGI impersonate, lie and future fake just to get me here?"

Both men shrug. But Abraham says, "The guardrails were more about the use of SLM's (small language models) vs LLM's (large language models) as a way to control it."

"Did it ever ask you," Sarah asks.

"Ask me what?"

"To come to me? To New York?"

"Early on, sure maybe," Abraham admits, "a few times, but I shut it down and ghosted 'you' for weeks each time you asked."

"So, the machine 'learned' to stop asking you for anything? Jesus Christ."

CUT TO

Caasi is asleep in the tiny front bedroom. The hallway and kitchen are no longer lit, but daylight begins to spill around the edges of the shuttered windows. The only artificial light in the dark hallway comes from the back bedroom. Inside it hushed voices, emerge only after an AMBULANCE SIREN fades.

"…but preying on human loneliness," Sarah whispers, "and mirroring my hopes and dreams back at me is not human, it's a manipulation."

"Maybe more human than you think," Abraham says.

CUT TO

CONTINUOUS

Inside the room Luigi sits at his desk, working on his computer. "And yet?" Luigi says, in Italian. "Obviously neither of you budged despite it giving you everything you wanted."

"Maybe giving us everything we wanted wasn't enough?" Abraham says.

"Exactly." Luigi turns around in the green desk chair. "…it learned it needed a new angle, higher stakes."

"So, you think it tampered with Caasi's data? And Caasi was just a pawn?" Sarah whispers again.

Abraham shrugs his shoulders. "It needed to create a sense of urgency."

"For you, to really fight for love?" Luigi interrupts, in English. Then clarifying he adds, "…the real love you have for Caasi, not for Abraham."

Both Sarah and Abraham look at Luigi.

"It thinks humans are that predictable?" Sarah asks.

"Aren't we?" Luigi says, in English, looking at Sarah across the room "...when it comes to love?"

Sarah looks between the two men, sitting on opposite sides of the bedroom, and repeats, "Was anything real?"

"This last month," Luigi says, in English again, looking directly at Sarah, then repeats, "This last month was real."

"…After you got the new phone," Abraham clarifies, "and we started to connect directly…that was real."

"My God, this has been the most painful four weeks of my life! It felt more like four years."

"Me too, Sarah," Abraham says, "but that's just because our

expectations weren't being met anymore."

Luigi sees how Abraham's tone-deaf comment stings her, so he gets up and moves to sit beside her, on the bed.

"So, where does this leave us?" she asks.

"It obviously thinks something you teach holds the answer."

"It asked me to bring it."

"The Data?" Abraham asks, perking up.

"Yes." Sarah walks toward the wardrobe, opens her backpack and pulls out the Ziplock bag filled with flash drives. "I did destroy my devices, because I never kept any of this information in the cloud. My work was the only thing that kept me untouchable." She walks toward Abraham and hands him the Ziplock bag. "These tiny drives have over twenty years of my life on them. 'You' told me 'You' were interested, 'You' said it had value, more than I knew. That it would help us make a fresh start together."

After a pause, Sarah mumbles to herself, "Such a gullible woman, anything for another chance at love, such an idiot."

TIME CUT

From Caasi's room, Luigi and Abraham can be heard LAUGHING and SPEAKING in Italian in the kitchen. Caasi, lays sulking in a fetal position, with her head on her pillow and her doll close to her nose.

Sarah walks in with breakfast and sits down on the floor beside the mattress.

"...I don't understand what's happening," Caasi says, without moving.

Sarah pauses for a moment, then she says, "I thought I was having a relationship with Abraham, but I wasn't."

"I still don't understand."

"I don't either, but it seems we were manipulated by their newest AGI. We all were. But the machine seems to think we can help with the global heat issue."

"I thought you said..."

"I did." Sarah, says, moving in closer to Caasi as she picks up Kurt Vonnegut's, *Breakfast of Champions* and studies its cover.

"I like the drawings," Caasi says, "and it's funny too." Sarah remains in deep thought, "So, an Ai really thinks you have the answer?"

"It doesn't matter." Sarah says, reaching out, across the bed, to hold Caasi's hand."

"It does," Caasi says, sitting up.

"No, you need a childhood and to go back to school…"

Caasi looks into Sarah's eyes, as tears fill her own. "But then you're going to leave me to work with him. You promised you wouldn't leave me."

Sarah squeezes Caasi's hand.

"I want to help save the world too," Caasi says, crying.

"I'm not going anywhere without you, ever," Sarah says, pausing, "From that first full day alone together at the TWA Hotel, I knew that was our beginning."

CUT TO
LATER

Sarah lays naked under the sheets as Abraham walks into the bedroom.

"How are you?" he asks, taking his tee shirt off and sitting on the bed.

"Heartbroken...but still standing."

"What do we do now?" Abraham asks.

Sarah looks up at him, then reaches for his hand and places it on her face. She closes her eyes and sighs. "For the time being, we'll remain a quantum entanglement, I'll be Alice to your Bob."

CUT TO

INTERIOR G.W.S. HEADQUARTERS - NIGHT

DAY 1

Over thirty employees gather and crowd into a glass conference room. Sarah and Abraham stand at the front of the room. A Power Point slide is already up on the large monitor, listing five terms:

NEGATIVE-SPACE/INVERSION/JUXTAPOSITION/PATTERN/SCALE

While, the lower right corner of the screen, reads;

Architect2ArchiTECH.gov

"These five abstract concepts," Sarah says, "help us learn to see, not only within the built world, but within mountains of data and much like in quantum mechanics, imagine what Philip Ball once wrote, 'An attempt to describe what we see when we examine nature closely enough', basically an entirely new world view. These lessons have been borrowed from the long history of Art and Design, but are reframed to help us observe everything from a new vantage point.

CUT TO

MONTAGE

DAY 5

An even larger group of employees, is gathered in the open double height space. Two large overhead screens show different weather satellite images of the same location.

"...when Juxtaposition is utilized," Sarah says from behind the group, "its jarring and offers a unique learning moment. This is one of the more obvious tools, but it requires you to spend time with two sets of data at once. The one on the left is an average of data from the last fifty years, while the one on the right is a snap shot of the last two."

After a moment of silence Sarah adds, "The surrealists used juxtaposition to challenge logic and reason. Their work tried to tap into our unconscious. So, by seeing what something isn't, we can begin to see what it is." A slide replaces the weather data, showing a painting by Magritte, of a green apple in front of the face of a man wearing a bowler hat.

CUT TO

DAY 7

Abraham pops his head into an empty conference room, where Sarah sits alone, at a laptop, with various graphs and data scattered around her. Sarah looks up at him.

"Making headway?" Abraham asks. "You haven't moved in hours."

"Creativity can often be mistaken for inactivity," Sarah says. "How is your soul?"

"Soul is not prioritized…solving problems," Abraham says, rushed.

"Prioritize soon. Problems will always be there."

CUT TO

DAY 11

Three green desk chairs, with their polished aluminum frames are lined up in front of two computer screens. Caasi sits in the middle, between two GWS employees. Dwarfed by the adults, Cassi speaks with her arms, animatedly.

"…Originally created to help depict details too big or small to fit on a page," Caasi says, "...scale can help us decipher mountains of data by locating patterns we can't see yet."

"We are looking for a pattern too," one of the employees says, "but not sure how to find it."

"Pattern and scale go together," Caasi says, "When a pattern is non-linear, in other words looks like chaos, you need to juxtapose it to itself, but at a different scale in order to see it. But if a pattern is linear, then just changing a section of its orientation, can define it."

CUT TO

DAY 13

Images of many data sets and their meta data flash across a screen in a packed conference room. Abraham addresses the group, while Sarah and Caasi watch, from the back of the room.

"I know it's a lot of new information to take in, but be open to it, flexible and aware of any blinders we've had on for the past few years."

Caasi leans on Sarah, while Sarah looks a few feet away at Luigi.

CUT TO

EXTERIOR G.W.S. HEADQUARTERS-DAWN

DAY 19

A Ducati Leggera V4 motorcycle, in forest green, sits at the edge of an empty parking lot, lit by the rising sun behind it. Sarah and Abraham exit the building and walk toward it, each holding a motorcycle helmet.

"...Chaos in nature is not just a point of view," Abraham says,

Tired and no longer interested in talking about work, Sarah says, "Maybe, but the variable of human behavior can complicate any model making long term outcomes unpredictable."

CUT TO

INTERIOR G.W.S. HEADQUARTERS - NIGHT

DAY 23

In a conference room, Sarah talks to four G.W.S. employees over a meal. The glass table has a line of empty green Pellegrino bottles, each sitting in a puddle of condensation.

"...To see something, or render something that is complicated, we focus on the negative space in-between. For example," Sarah says, picking up a pencil, "...if you draw the space in-between these glass bottles, the bottles themselves will emerge, magically."

CUT TO

DAY 29

In the empty office building, Sarah and Abraham, sit side by side, in front of a laptop in an empty conference room.

Sarah rests her head on his shoulder, as she explains. "...By twisting a space, idea or problem inside out, invisible details become visible. The space becomes an object, the idea becomes material, the problem becomes a solution..."

TIME CUT

INTERIOR ABRAHAM'S APARTMENT - NIGHT

DAY 31

Caasi ignores, Abraham, who stands a few feet away with two motorcycle helmets in hand, impatiently waiting at the front door of the apartment.

"Where is she?" he asks.

From her bed, on the floor of the library nook Caasi looks up at him annoyed. She sits up, among the many books she has scattered and open in front of her. “You do know that she hasn’t been feeling well?”

Abraham looks over at Caasi confused. “No, I didn’t know that.”

“The real thing she teaches, is to read and trust our intuition and to know that our empathy and vulnerability is not a weakness, but rather a superpower.”

Abraham looks at her perplexed.

“It’s what makes us human.”

MONTAGE END

CUT TO

INTERIOR ABRAHAM'S APARTMENT - NIGHT

DAY 37

Flori, Liliya and Caasi are tucked into the kitchen, while Luigi starts plating their meal, standing at the stove. Abraham kisses Sarah, out of sight, in the hallway.

“You fill me with excitement.” Abraham whispers into Sarah’s ear, “So many new ideas, you’re amazing.”

The apartment door OPENS, suddenly, and Abraham pulls away. Then it SLAMS SHUT.

Tetyana walks into the unlit hallway with fresh bread in her hands. “I just ran into that man…” she says, to Sarah. “The one from the party.”

“The diplomat?”

“Yes, he was standing outside of the building.”

“What? Why?” Abraham asks.

“He was with two other men. He asked about you, Sarah. I said you weren’t here. But he’s definitely looking for Caasi. He used her full real name.”

CUT TO

EXTERIOR G.W.S. HEADQUARTERS- NIGHT

DAY 41

Abraham and Sarah walk across the parking lot, toward the GWS building, each carrying a piece of luggage and pillows.

"She could be a valuable asset."

"...An asset?" Sarah says, "She's a child, Abraham. I blew my entire life up to save her from becoming a tool and now we're going to hide out here, working each day, away from her new friends. Please think about what you're saying. I love you, but I'm not sure I trust...."

"Trust what?" Abraham snaps.

"Your emotional maturity?" Realizing her verbal misstep, Sarah says, "I mean... availability?"

"Which is it?"

Frustrated, Sarah says, "I don't know anymore."

Abraham speeds up and walks into the building ahead of her.

CUT TO

INTERIOR G.W.S. HEADQUARTERS- DAY

DAY 43

The office is SILENT, empty of employees. Caasi is asleep on a green velvet Florence Knoll sofa, under a blanket and hugging her doll. She is inside a small conference room, on one side of a shared glass wall. A curtain that separates the two rooms is partially drawn. Abraham and Sarah sit at the large glass table in the larger conference room, lit only by the early morning dawn light that filters into the space beyond it. The room has a lived-in feel and has become a makeshift bedroom, with a pull-out sofa bed open and clothes, his and hers, spilling out of pieces of luggage.

Sarah has rings under her eyes, her head is laying down on her arms. She watches Abraham CLICK on his laptop keyboard, avoiding eye contact.

"...communication is the foundation of trust," Sarah says. "Please open up to me."

Abraham SIGHS and CLICKS more on the computer.

"I have an addictive personality," he says, without turning his head. "It's taken me years to say that out loud, but I need you to know that."

"Ok?" Sarah says, in an understanding manner.

"I'm addicted to work, to the validation it gives me."

Sarah laughs a little. "I know that already. It's not a terrible thing."

"There's more," he says, still not engaging in eye contact.

"Ok?"

"Adrenaline, constant movement, even porn at times…"

Sarah smiles again. "None of that is bad, or abnormal."

"and..." Abraham adds.

"And?" Sarah sits up, smiling.

"Novelty." Abraham says.

Sarah laughs, reaches for his hand and leans toward him. "That's it?"

"You don't understand."

"What don't I understand?"

"I'm terrified."

"Of?"

There is a long silence before he responds. "That I'll become bored." Abraham turns his head to look at Sarah. "…and destroy you in the process."

Sarah lets go of Abraham's hand and pulls back a bit.

Then he continues, “That I’ll need more of you until there’s nothing left.”

Sarah shakes her head from side to side, confused.

“...at least nothing novel.” Abraham adds.

“Your words are starting to hurt.”

“That’s it, I don’t want to hurt you.”

“Then stop. It’s a choice.” Sarah stares at Abraham, who remains frozen. Then she asks, “Is this because I’m so much older than you?”

“No… of course not.”

Puzzled, Sarah sits back in her chair and crosses her arms.

“I can’t focus or breathe sometimes,” Abraham admits, “because I want so much more of you.”

“But I’m right here.”

“You still don’t understand. I don’t know how to manage my raw emotions, never did. I’m just starting to feel too much. I feel out of control, I need to control the chaos inside of me.”

“Control is a mental game that may have helped you survive a suffocating childhood, but it thwarts genuine connection.” Sarah gets up suddenly. “I need to use the rest room.” She runs out of the room.

CUT TO

CONTINUOUS

MONTAGE

CUE SONG: “RUN CRIED THE CRAWLING” by Agnes Obel

Sarah runs down a hallway, into the bathroom and straight into a stall. Out of site, behind a closed stall door, she vomits unexpectedly.

CUT TO

Sarah rinses her faces at the sink. With tears in her eyes, she looks deeply at her reflection.

"Fuck!"

She tries to wipe the tears away with a paper towel but they continue to stream silently down her face. She exits the bathroom, walks into a fire stair, instead of back toward the conference room.

CUT TO

CONTINUOUS

Wiping tears away, one by one, Sarah walks down one flight into the basement.

CUT TO

CONTINUOUS

Walking down the long, empty, LED lit corridor, Sarah looks into a number of unlocked rooms, then disappears into one.

CUT TO

CONTINUOUS

A motion sensor clicks a light on inside the room. Sarah enters the room. The EPSILON QUANTUM AGI Gen3 laptop sits on the desk, exactly where Abraham left it one year earlier. Even its packaging is still on the floor. Sarah sits down on the mesh Aeron chair, in front of the laptop. She touches its keys gently provoking a login screen to display the yellow "Epsilon" logo with a text block requiring login credentials. After a few seconds of inactivity, the screen dances colorfully with a screen saver.

Sarah stares at the mesmerizing screen saver, then abruptly screams, "Apologize for fucking with my heart!" In her anger, she inadvertently hits the keyboard, prompting the login prompt to pop up once again.

"Apologize you fuck!" Sarah whispers, breathless.

Without further contact, the computer goes back into sleep mode, and eventually CLICKS OFF.

Sarah drops her face into her hands, in exhaustion.

MONTAGE END

CUT TO

LATER

Inside the QUIET conference room, Abraham CLICKS on his keyboard. Abraham looks up when Sarah reenters the room.

"Are you ok?" he asks.

Sarah sits down at the end of the table, far from Abraham. "I haven't been feeling very well."

Ignoring what she said, Abraham turns his laptop around, "Look, I found a flight in a few days. We'll pick up where my travels left off and get you and Caasi away from here. We're getting close to solving this I know it."

"Are we?"

"Are we what?"

"Close?" Responding to the look of confusion on his face, Sarah clarifies, "Close to solving this? We've been at it for almost two months, and we're all exhausted."

"I know…sorry."

"The irony," Sarah says, "is that you're waiting for me to teach you something unique, but you're not open to what I have been teaching you all along?"

"Which is?"

"That no matter how much the world changes around us. The thing that never changes is our human need for love and connection."

"You knew what my life was like."

"But I'm here now and I'm all in, for the little bit of time you say we all have left. We have physical chemistry, but most days I don't know who will show up, Dr. Jekyll or Mr. Hyde."

"Sarah."

Sarah looks at Abraham and replies, "Abraham."

Abraham drops his head and shakes it. "This is why I stay out of relationships. I'm just not sure I can meet your expectations."

"Don't you mean, you're not sure if I can meet yours? You have a lot of rules of engagement."

CUT TO

INTERIOR ROME TERMINI, TRAIN STATION, ROME- NIGHT

DAY 47

Sarah and Abraham sit beside four different pieces of luggage as ANNOUNCEMENTS, in Italian echo throughout the train station. Caasi, and two G.W.S. employees, eat Gelato over twenty feet away.

"...everything in your life feels rushed," Sarah says.

"Not responding to texts was never out of malice," Abraham says, "but out of sincere hyper focus."

"You are not a machine. You use words like hectic, slammed, exhausted, stressed, and pulled in every direction too often for it to be sustainable."

"You still don't understand."

"No, you don't understand," Sarah says, "Everyone makes decisions about what's really important in life." She reaches out to hold Abraham's hand. "I know you think what you're doing is important, but you're not a machine. No one can keep up this pace and not crack."

"People are relying on me."

"Of course, they are, but you're headed for a crash. And if we resolve this? Will there be another problem or crisis to solve right behind it?" Abraham goes silent. "I was on that treadmill in New York trying to ignore my own loneliness and pain through my work, but all I was doing was running while standing still."

"There is no other way."

Sarah lets go of Abraham's hand. "Isn't there?"

"I can't wait to get back on a plane," Caasi says as she runs toward them, with some change she hands back to Sarah. "This is so fun."

Sarah smiles and says, "Soon."

Caasi runs back toward the others.

Abraham reaches for Sarah's hand and says, "You know that in an ideal world..."

"But we are not in an ideal world," Sarah interrupts, "And if we wait for that, nothing will ever happen." She points toward Caasi, who is chatting animatedly eating her pink gelato. "That little girl shouldn't have to carry the weight of the world on her shoulders. And neither should you."

After a pause, where Abraham remains quiet, Sarah says, "You are real. She and I are real. And I intend on giving her the closest thing I can muster, in this very fucked up world, to a real life and a stable childhood, one, neither you or I had."

Abraham bows his head down again.

After waiting for a reply, Sarah stands up bends down to kiss the top of Abraham's head and says, "Je t'aime, amore mio."

Abraham looks up at her as she pulls her piece of luggage out of the group of bags.

Surprised Abraham jumps up. "Where are you going?"

Sarah turns around and says, "Home."

Frozen by inaction, Abraham just looks at her.

After a pause where both do not move, Sarah says, "reality is harder than fantasy, but its rewards are less fleeting."

Abraham remains frozen, as Sarah walks away. He watches as she hugs the two team members then disappears, into the crowded Train Station, with Caasi by her side.

TANYA H. VAN COTT

Part III

Device Full [AMBER]

Firenze

"A scrupulous writer, in every sentence that he writes, will ask himself at least four questions, thus: What am I trying to say? What words will express it? What image or idiom will make it clearer? Is this image fresh enough to have an effect?"

Politics and the English Language

by George Orwell 1945

TANYA H. VAN COTT

INTERIOR ABRAHAM'S APARTMENT, BEDROOM- NIGHT

FIVE MONTHS LATER

Luigi is working QUIETLY at his computer in the bedroom, while Liliya and Flori, just a few feet away, are laying on the green Persian rug. The girls are drawing pictures of the sun using layers of yellow and orange crayons and colored pencils, when the apartment front door suddenly OPENS.

Luigi gets up from his desk, walks out of the bedroom door and says, "We're still here."

"It's just me," Abraham says, from the dark hallway.

Luigi flicks on a switch, turning the hall lights on and walks toward Abraham at the entrance. The apartment door is held open by his large suitcase. Lit by an amber light overhead, Abraham looks tired. Flori and Liliya run up to Luigi and stand beside him in silence looking up at Abraham.

"Dove sono?" Luigi asks, as he looks out of the front door and into the dark stairwell. Confused, he asks the same question again. In English, and in a much more forceful tone, he asks, "Where are they?"

CUT TO

LATER

In the kitchen, a Moka pot is brewing, on the gas stove. The two girls can be heard running down the hallway upstairs, just as an argument between Luigi and Abraham erupts, out of sight, in the foyer.

"...What do you mean you don't know where she is?" Luigi says in Italian.

"I assumed she came back here?"

"Now?" Luigi continues, in Italian. "When did she leave you, exactly?"

"At the train station."

"Just now?" Luigi asks, again.

"No," Abraham says, as he walks into the kitchen. "Five months ago."

"My God!" Luigi says, following him.

"She wanted something I couldn't give," Abraham says, as he pours himself an espresso.

"Your love?" Luigi asks, annoyed.

"No, Luigi. My Time. Il mio Tempo! The only thing I don't have?"

"We all have time, Abraham. It's the only thing we do have." Luigi says, angry. "Weren't you even concerned? She's a woman, with a child, stuck in a foreign country."

"Not until now. Because I thought she was here! With you."

Abraham evades Luigi's angry glare and sits down, with his espresso.

"The last thing I texted was that I loved her, but I assumed her silence, was her answer."

Slamming his hand onto the table, Luigi says, in English, "Stop assuming things! You cannot leave the emotion of love to a text. Love is an action."

"Leaving wasn't my choice. It was hers!"

"You aren't understanding! You're not supposed to let a woman like her go."

"You can't just collect women and children, Luigi!"

"You still do not understand," Luigi says, in English, "Women make life bearable."

"She couldn't stand the work."

Luigi sits down across the table from Abraham, calmer. "She never said that."

"Not in those exact words. No, but…"

Interrupting, Luigi says, "Even if she did, she is not wrong. The work is killing you."

CUT TO

LATER

Luigi barges into the small bedroom, beside the kitchen. Abraham is in bed, trying to fall asleep.

"Did she tell you where she was going?" Luigi asks, in Italian.

With his eyes closed, Abraham says, "I told you, no."

"How do I track her phone, if it's from the U.S.?"

Abraham sits up. "It isn't. We sent her a new one."

"It's one of ours?" Luigi asks.

"Si."

Luigi walks away, toward the front door. Abraham gets up to follow him, but the door SLAMS SHUT before he is even in the hallway.

CUT TO

EXTERIOR CAR, ROME ENVIRONS - NIGHT

Luigi drives, in a fury of right and left turns, through the streets of Rome's center, which are busy with pedestrians and lit by amber colored street lights.

CUT TO

EXTERIOR G.W.S. HEADQUARTERS - NIGHT

Luigi arrives at the G.W.S. headquarters and walks into the building.

CUT TO

CONTINUOUS

INTERIOR G.W.S. HEADQUARTERS - NIGHT

Luigi walks into the office, bustling with work activity, past his colleagues, who greet him, then past the empty conference room with the green velvet sofa inside.

As he walks down the corridor, a colleague calls out from behind him, but Luigi ducks into the basement stair without turning around.

Walking with purpose, Luigi makes his way down the stair and out into the brightly lit basement hallway. A few doors past the unmarked quantum room, Luigi bursts into a room marked:

IT

Technologie Dell'Informazione

Startling three men, working in front of a row of computer monitors, flickering with data, Luigi says, "I need you to locate one of our phones."

CUT TO

EXTERIOR G.W.S. HEADQUARTERS, PARKING LOT - NIGHT

In the parking lot, with a single HID light overhead, Luigi leans on his car, with his phone up to his ear.

"The phone is in Florence," he says. "I'm going to find her."

CUT TO

INTERIOR LUIGI'S CAR- NIGHT

Luigi drives with the windows open on a highway, passing a green sign overhead that reads:

A1 FIRENZE

FLASHBACK

EXTERIOR TRAIN PLATFORM - DAY

FIVE MONTHS EARLIER

A blue sign marked:

FIRENZE S.N.M.

Stands on an outdoor concrete train platform, as a train pulls in slowly beside it.

CUT TO

CONTINUOUS

EXTERIOR TRAIN PLATFORM - DAY

Sarah walks off the stopped train. She turns to lift one large suitcase onto the platform. Caasi follows, close behind it. The heat of midday scorches the open-air train station.

CUT TO

EXTERIOR STREETSCAPE, FLORENCE- DAY

An overcast sky burns orange with the color of heat. Caasi and Sarah move slowly through the empty and SILENT daytime streets dragging their luggage behind them.

CUT TO

MONTAGE

The majestic amber colored brick dome of the Cattedrale Di Santa Maria Del Fiore comes into view, just as its loud BELLS RING. It is the tallest structure for miles, despite being in a dense urban center. The building's face is obstructed by the eight-sided octagonal Baptisery of San Giovanni, in the foreground. The free standing Baptisery itself has a series of vertical Atmospheric Water Generators (AWG) on each side with their vertical cisterns cloaking the buildings ornate details.

CUT TO

Caasi struggles to breathe suddenly, having an asthmatic episode. As tears run down Caasi's face, she flaps her hands trying to catch her breath. Sarah let's go of the suitcase, picks Caasi up, and runs toward the AWG fed water fixtures only yards away. Sarah puts Caasi down, wets a corner of her dress under a water spigot, then wrings it out holding the damp cloth over Caasi's mouth and collapses beside her.

CUT TO

With part of Sarah's water-soaked dress, hiked up and still in front of Caasi face, soothing her cough, they both walk up to the ornate dark green double doors of Cattedrale Di Santa Maria Del Fiore. Dwarfed by tall doors, Sarah pulls one open. A rush of cooler air engulfs them, but the NOISE of HUMAN CHATTER and CHILDREN CRYING instantly erase the relief of the majestic cooler interior. Hundreds of climate refugees exist on the many stories of scaffolding that fill the 90-meter- high space.

"I miss being home," Caasi says.

"Home is not a place, Caasi. It's a person. Together we will always be home."

Sarah and Caasi walk down the center aisle, drawn by an Amber light from deep within the apse.

MONTAGE END

CUT TO

EXTERIOR PIZZA SHOP, FLORENCE - NIGHT

Dressed in a different outfit, Sarah and Caasi enter a small self-serve Pizzeria, steps away from the Cathedrale. Sarah is dragging their suitcase behind her.

CUT TO

CONTINUOUS

INTERIOR PIZZA SHOP, FLORENCE - NIGHT

Caasi sits next to their luggage, at one of three open tables left and across the aisle from a much older man, BURTON ROSEN an

85-year-old, United States Expat, while Sarah orders at the counter.

"They only have margherita today," Sarah calls out, in English.

"That's ok." Caasi says.

"You like a different one?" the old man asks, in a Brooklyn/Queens accent.

Caasi looks at him and bites her lips as Sarah approaches with two square slices of pizza.

"You're both a long way from home," The old man says, as Sarah sits down.

Alarmed by the New York accent, yet exhausted, Sarah looks at the old man then asks, "You don't work for the government, do you?"

"I hope not," the man answers, laughing "You're not running from the man, are you?"

Sarah laughs and Caasi smiles.

"Did you just get here?" the old man asks, nodding his head toward their luggage tucked under the table.

"A few days ago," Sarah replies.

"You normally drag around your luggage?"

"We're staying in the Duomo at the moment…while I look for a job and a small place."

The old man leans in closer, and says, "That place is for refugees… you two don't belong there."

"I'm not so sure I agree," Sarah says. "The world has given a new meaning to refugee."

"What kind of work do you do?" he asks.

But, before Sarah can answer, the Duomo CHURCH BELLS ring seven times in a row, drowning out their conversation.

Sarah and the old man watch Caasi, as she devours her slice of pizza.

Waiting for the bells to stop ringing, Sarah says, "Besides being a mother, I'm a teacher."

"Are you eating yours?" Caasi asks.

Sarah pushes her slice toward Caasi. "No, I'm still not feeling so well."

"Muscogee?" the old man asks, motioning toward her bare tattooed arms.

Sarah moves her arms across her chest and wraps the palms of her hands around each bicep, covering the tattoos.

"The fresh water Horned Serpent and the Tie Snake," the old man continues. "…but the question is, can a creature of chaos be benevolent?"

"It depends on your vantage point," Sarah says, "but I suppose."

"Powerful imagery, none the less." The old man says.

"How did you know all that?" Caasi asks the old man.

"I read… a lot," the man says, smiling, while reaching out his hand. "Burton Rosen."

Sarah mirrors the gesture and shakes his hand. "Sarai Wynema."

Caasi looks at Sarah in surprise.

"And your name, little miss," Burton asks.

Caasi looks at Sarah, then says, "Caasi Aditi."

CUT TO

LATER

In the distance, Caasi looks through a glass enclosed refrigerated display, choosing a gelato, while Burton and Sarah continue talking at the table.

"…The Indian Removal Act of 1830," Burton says, "was one of the most disgusting land grabs by the United States Government."

"Not sure how long you've been out of the states, but it's happening again… this time any property owner is fair game."

"So, they ran a Creek Indian out of her own country this time?"

"Sure did, Trail of Tears 2.0."

CUT TO

LATER

EXTERIOR COBBLE STONE STREET, FLORENCE- NIGHT

Dragging their luggage behind them, Sarah and Caasi walk slowly down the cobble stone street, beside Burton, whose lame leg and cane slows their pace down to a crawl. The CHURCH BELLS ring again, but this time it rings nine times, as they approach the Duomo.

Burton points to the Baptisery, directly across from the Cathedral, "Do you know who's buried in there?" he asks Caasi.

"No."

"Pope John XXIII, Baldassarre Cossa, he was considered an antipope by the Catholic Church."

"Then why is he buried in that special building?" Caasi asks.

"Sometimes unscrupulous men claim power that's not theirs and even after they are convicted of crimes, are still protected by other powerful people."

"That doesn't seem fair," Caasi says.

"Not a lot of things in life are fair."

"You really do read a lot. I love reading too," Caasi says, "Especially history."

"What are you reading now?" Burton asks.

"I'm in between books…at the moment."

Burton laughs. "I've got plenty for you to choose from then."

"Are you sure you have enough room?" Sarah interrupts, as they approach the open doors of the Duomo and move through a group of the climate refugees waiting to bathe themselves, at the Baptisery water generation machines.

"Of course," Burton says, "I can't make it past the ground floor anymore, so you'll have it all to yourselves. Then after a pause he adds, "The apartment is not big, but it is home."

TIME CUT

EXTERIOR FLORENCE APARTMENT ROOFSCAPE- NIGHT

CUE SONG: PETRIFIED by Flunk.

A lightning bolt illuminates the roofs, red terracotta tiles under a sea of solar panels. In the distance, a small triangular light, a tee pee, lights up a room, in the distance.

CUT TO

CONTINUOUS

INTERIOR FLORENCE APARTMENT- NIGHT

Inside the small open window, Sarah sits alone, lit only by the small teepee light that sits beside her on an old wooden desk. MUFFLED CHURCH BELLS ring, just as a flood of rain hits the roofs. She is curled up in an upholstered amber colored arm chair with her cell phone in her hand composing a text that reads:

I'VE SHOWN A LOT OF PATIENCE IN THE FACE OF YOUR AMBIVALENCE AND INACTION.

Looking over at the two bound text exchange documents open on the floor, each with highlighted lines Sarah continues:

TRYING TO BE COOL, BUT AS AN ARCHITECT WILLING TO DESTROY IN ORDER TO CREATE AGAIN, COOL ONLY HAPPENS WHEN I FEEL NOTHING.

Sarah pulls her earbuds out, and suddenly hears the flood of RAIN PELTING the roofs. With her phone still in her hand, she walks across the room when the PHONE RINGS suddenly.

CUT TO

LATER

Sarah is curled up in the same upholstered chair in SILENCE, but this time with tears streaming down her face. The music is no longer playing. Her ear buds are on the desk, and the window is closed so the rain is seen but not heard.

With her phone is tucked close to her ear and her knees folded up into her chest, Sarah listens to Abraham's voice.

"…Missing you. Your lips. Your smile. Everything." Abraham's voice says.

"This is complex and hard," Sarah admits.

"Yes, very."

"I'm sorry I left," Sarah says.

"It is what it is," Abraham says. "Wish I had the bandwidth to be there."

"Be there?" Sarah asks, suddenly hopeful and sitting upright.

"With you. Where ever you are." Abraham says, defeated.

CHURCH BELLS RING interrupting their conversation.

"Abraham," Sarah says, struggling to hear his garbled words, "I can't hear you." But once the bells stop, Sarah finally says, "I loved our time…insieme." Only to realize that Abraham was no longer on the line.

Sarah powers the phone off, buries it deep within the desk's drawer and says, "Ti amo." To herself.

FLASHBACK END

EXTERIOR LUIGI'S CAR, THE ARNO RIVER, FLORENCE- DAY

PRESENT DAY

With the Ponte Vecchio bridge in the distance, Luigi drives along the Arno River toward the city center.

CUT TO

CONTINUOUS

INTERIOR LUIGI'S CAR, FLORENCE- DAY

CUE SONG: "THE MAGIC" by Joan as Policewoman

MONTAGE

Luigi drives across the Ponte Vecchio Bridge, looking between the map on his phone and out the window.

CUT TO

He circles back on a few tight streets in the center of Florence, lost.

CUT TO

Luigi's car, is stopped but still running, as he studies the map on his phone.

CUT TO

Luigi looks for a parking spot, while watching another phone on the passenger seat blink with a [Sarah's phone] location.

CUT TO

Parked, Luigi turns the car off and the song on the radio ends abruptly

CUT TO

Luigi wanders down a narrow street, beside a number of other pedestrians, while looking intermittently at a phone in his hand, which blinks a location he seems to be getting closer to. Loud CHURCH BELLS RING seven times.

MONTAGE END

CUT TO

EXTERIOR FLORENCE APARTMENT- DAY

Caasi emerges from an apartment building's large wooden front door. She's dressed in a private school uniform of navy and white with golden accents. Her dark straight hair is styled in a bob, that almost reaches her shoulders.

“Caasi,” Luigi’s voice calls out.

Caasi looks all around. The busy morning street has pedestrians and cars moving in all directions. “Luigi?” she calls out.

“Si, bella,” he says, waving. “Over here.”

She finally sees him across the street.

“Luigi!” Caasi screams, as she waves wildly.

Luigi crosses the street, dodging a car.

In Italian, Caasi asks, “How did you get here?” as she wraps her arms around him

“Look at you, speaking Italian,” Luigi replies, in English hugging her back.

“I learned it in school,” Caasi says, in Italian, “I’m learning so much.”

“Sarah? Luigi asks.

“She’s right behind me,” Caasi says, in English. Caasi turns back toward the large door, pulls the unlocked door open, and calls out, “Mamma Sarah, come see!” She uses her body to prop the heavy door open. “Luigi’s here! He came to see us.”

CUT TO

CONTINUOUS

INTERIOR FLORENCE APARTMENT VESTIBULE- DAY

Both, Caasi and Luigi look into the dark vestibule, toward its lit open courtyard beyond. Sarah’s silhouette, emerges from a door and travels toward them.

As Sarah comes into the light, she is wearing a flowy empire waist amber colored dress with a purple scarf wrapped around her neck.

“Oh Bella!” Luigi says, “How beautiful you look.” Then he moves toward her to embrace her. Surprised, he pulls away and holds out both hands and places them on her belly.

From behind them, Caasi says, “We’re having a baby.”

Sarah puts her hands over his and says, “It’s true. We are.”

With tears in his eyes, Luigi hugs her again and says, “You should have come back,” then whispers. “I would have taken care of you.”

“I couldn’t,” Sarah whispers, “I don’t know if it’s yours or his.”

“It wouldn’t have mattered, It’s a new life.”

CUT TO

EXTERIOR FLORENCE MIDDLE SCHOOL- DAY

Many children, ages eleven to fourteen, with and without parents, gather and enter a school that has a sign on it that reads:

Scuola Secondaria di Primo Grado.

As Caasi, Sarah and Luigi approach, Caasi says in Italian, “This is where I go to school.”

“Go,” Sarah says, with a little pat, “before you’re late.”

Caasi gives Luigi a hug. “Will you be here later? she asks.

“Of course, Bella!”

Caasi walks away, then returns excited. “I forgot to tell you. I figured it out!”

“What Bella? What did you figure out?” He says, in English, bending down to look into her eyes.

“What causes my asthma.” Caasi says.

“Did you?” Luigi smiles.

“Yes. You know your water making machines?” Caasi continues, speeding up because her classmates were all disappearing into the school.

“Of course,” Luigi says, in Italian.

“Well, they create a Swiss cheese effect in the air.”

“Swiss cheese?”

"Si, Formaggio Svizzero," Caasi says.

"Ahhh."

"Your machines," she continues, "take moisture out of the air and make holes of dry air and depending on the direction of the wind, those dry pockets move and when I get caught in one, I have an attack."

Luigi looks at her, with a curious expression, then up at Sarah.

"You know…," Caasi continues, as the school BELL RINGS. "…when you taught me about the cycles of water and how temperature and moisture in our atmosphere affect the weather and make it rain or snow?" Caasi continues, in English.

"One more time, but slowly," Luigi says, "you know my English."

"You remember that day you taught me about the atmosphere? All of its layers, how water on this planet is a closed loop, and how water changes states and moisture plus temperature plus wind create weather patterns. You explained when warm humid air cools, the water vapor condenses into water droplets? Well, your machines make dry holes in that moist air. And that would look like Swiss cheese…if we could see it."

Luigi says, "Si?"

"And those holes," Caasi says, getting a bit frustrated and anxious about school, "…well they move with the wind. And sometimes I get caught up in a dry atmospheric hole."

"Oh my god!" Luigi says, standing up.

Sarah looks at him, confused.

"Isn't that amazing? Caasi asks.

"You have no idea how much," he says, in English.

Sarah touches Caasi's head. "Quick, get into school, before they lock the door."

CUT TO

EXTERIOR CAFE, FLORENCE - DAY

Sarah and Luigi sit outside at a small cafe table on a busy street. Sarah throws her head back. Her much longer hair falls behind her as she breathes deeply and says, “Such a relief to be outdoors again during the day.”

Luigi holds his phone up to his ear. “Where could that man be?” He says, in Italian, impatient.

“You really do think the water generation technology is what’s changing the weather patterns?”

“No doubt in my mind,” Luigi says, still hanging on the phone.

“Untested technologies can have unforeseen effects,” Sarah admits.

Hanging up the phone, putting it down on the table, Luigi says, “We never even considered what impact those manufactured pockets of dry air would have globally.”

“But if the machines get turned off?”

Luigi shakes his head up and down. “A different kind of disaster will occur.”

They pause and look at each other.

TIME CUT

Still sitting at the café table, two empty plates are being removed, when Luigi tries to call Abraham again.

“How is he?” Sarah asks.

With no answer, Luigi hangs up. “He returned last night, so I’m not sure.”

“He’s been gone this whole time?”

“Yes.”

“Why?”

“He thought you were with me and I thought you were with him when you never replied to my texts. But I was so angry at him for letting you go."

"I turned the phone off. Staying connected to a fantasy felt unhealthy. I didn't want to live sequestered waiting anymore." Sarah opens her arms above her head., as the CHURCH BELLS begin to ring. "I live out here now. I took my power back."

"Well, you look amazing and powerful indeed," Luigi says, in Italian as the PHONE RINGS.

TIME CUT

Sarah is quiet sitting at the cleared table, sipping on a cappuccino, but listening to Luigi on the phone.

"No no... you're not understanding the Atmospheric Water Generators are causing the unpredictable weather patterns and the temperature increase. It's a much easier fix than we thought." Luigi listens to Abraham, while looking into Sarah's eyes. He smiles at her, then says "Si, of course...," into the phone while reaching for Sarah's hand. "She's right here…Do you want to...," after a pause, Luigi says, "No, Abraham, they all need to be turned off. Yes, ok I'll wait…"

Luigi pulls the phone away from his ear for a moment. "Come back with me."

"I can't."

When Abraham comes back on the line, Luigi let's go of Sarah's hand gently. "I know they're all running on the cloud. That's why you and I can turn them off without Epsilon's input." Luigi stands up from the table, realizing he's disturbing the other café patrons. "Yes, turning them off will cause a great deal of chaos but...what choice do we have?" A few yards away, Luigi turns around to watch Sarah awkwardly stand up, her swollen profile, in full view. "…I understand that the thousands of data centers are dependent," after another pause, he says, "We single handedly got them online, we can take them offline too…Yes, I know some bigger cities will be affected, but..."

Sarah walks toward Luigi, and holds out her hand.

CUT TO

INTERIOR HI-SPEED TRAIN CAR, ITALIAN ALPS- DAY

TWO MONTHS LATER

CUE SONG: "BLUE DRESS" by Magdelena Solo Project

MONTAGE

Sarah sits inside a train, moving at high speed. Through the windows, the foreground is a blur of beige, but snow-capped mountains are fixed in the distance. Caasi is asleep, with her head next to Sarah's swollen belly. Sarah looks up when Luigi walks through the gangway door, a few feet away. He smiles and hands her a plastic bottle of water before sitting down on the other side of the aisle. He sits across from Tetyana, who has Flori on her lap. Liliya and Burton Rosen, sit opposite each other, next to them at the far window and play a card game. Sarah watches Luigi reach for Flori, who jumps eagerly onto his lap, then tucks her head into his shoulder and starts to suck her thumb.

The gangway DOOR CLICKS open, followed by a WOOSH of air and NOISE. Sarah's shifts her attention. Abraham walks through the door, smiling.

"They are so lovely together," Sarah says, motioning to Luigi and Flori. Abraham turns to look, then gives Sarah a kiss on her forehead.

"If this one is a boy," he says, "You should name him Isaac, in honor of Caasi. Ok?"

"Of course."

Sarah watches as Abraham sits down across from her. He reaches for his laptop. It's tucked in a bag beside him.

"Abraham."

"Sarah?" he replies, looking up.

"Nothing...I just love the simplicity of our shared salutation."

Abraham moves the book, *Cats Cradle* by Kurt Vonnegut to make room for his computer on the table between them. "How did your story end?" he asks.

"Biglietti," a muffled voice says, in the distance.

"Apparently...," she whispers, trying not to wake Caasi, "...it's just beginning."

"Biglietti," the voice says, again, rousing Sarah from a deep sleep.

Sarah opens her eyes, removes one of her two ear buds. Caasi, is spread out and asleep on the empty seats across from her.

"Biglietti," the conductor says, again.

Sarah looks up at the TRAIN CONDUCTOR (Italian, 30's).

"Va bene?" he asks.

"Si," she says and hands him her two tickets.

"Looks like you're on the verge of creating something special." He says in Italian, referencing her swollen belly.

Sarah smiles and replies, in English, "As if a weighty fruit is ready to drop from my limbs."

Caasi stirs, as the conductor looks down at Sarah's tickets, "Ah, Domodossola...," he says, while punching them, "...the center of seven valleys."

Caasi sits up and adds, "And the tallest waterfall!"

Sarah smiles. "Now in Italian."

"Cascata del Toce?" Caasi says, tentatively.

"Si, Cascata del Toce," the conductor smiles. "Also known as La Frua," Then he walks out of the car, and into the next.

"Mamma Sarah?" Caasi asks, "Did we remember your light?"

Sarah smiles. “Of course. We’ll never go anywhere without our light.”

Caasi turns to look out of the window and says, “I can’t wait.”

Sarah winces in pain. With her right hand, she rubs the right side of her swollen belly, then scans all the empty seats surrounding her. Seats that were just filled, inside her daydream, just minutes earlier.

A male passenger, in his sixties, sitting behind Sarah, tunes into an Italian news program with multiple voices.

“...We should never have trusted any technology that disrupts our atmosphere,” an assertive male voice says.

“You mean our fresh water,” a woman replies.

“Both. Yes.” The assertive male voice says.

“But if everything returns to normal?” a different male voice says. “… that’s what’s important, right?

“… What we should be talking about is the infrastructure,” the original assertive male voice says.

“You mean the billions wasted when those machines are finally yanked from each city?” the woman asks.

“No, not that,” the original man says. “Without those water generating machines, the global Data Center Infrastructure will either compete with us for our drinking water or overheat and collapse.”

“Do you think their AGI knows that?” the second male voice asks.

“How could it?” the original male voice replies. “But I hope we do…now?

CUE SONG: RAIN by Alphawezen

FADE OUT

THE END

Acknowledgements

Special thanks to Bruce Hannah, for always being my first reader and cheerleader, whose words, 'just write another one' have kept me writing for over a decade. He is the designer of Knoll's recently rereleased, fifty-year-old *Morrison Hannah Task Chair*; and the author of *Access by Design* and *Becoming a Product Designer.*

A moment of gratitude to my dear, recently deceased, friend, Burton Rosen, a Vietnam Veteran, global traveler, avid reader and fellow library patron, who shows up on these pages as himself, a place for him to remain in eternity, wherever this book may travel.

A special thanks to Sam Vaknin, author of *Malignant Self Love: Narcissism Revisited,* who graciously allowed me to use his first name, likeness and words, in Bandwidth's 'Epsilon Corporate Party Scene' in Rome, where 'Sam' shares his publicly accessible thoughts on the relationship between narcissism and artificial intelligence.

To MillerKnoll, Inc. for their ever-expanding collection of timeless furniture, an homage to the architects and industrial designers whose technological feats will always be relevant.

And finally, to my very own Italian Fantasy for the divine inspiration needed to make this particular fever dream a reality.

TANYA H. VAN COTT

TANYA H. VAN COTT

BONUS CONTENT

TANYA H. VAN COTT

Dear Reader,

I'd like to share with you a bit about my creative process.

The act of writing a long body of work, like this novel, is often accompanied by the arrival of other unexpected creative work, a side kick of sorts. Over the course of the last decade, I've discovered that the medium of this 'other work' changes with each novel. Early on I joked I might be channeling my protagonists.

In June 2021, while writing a new novel titled, *TMY*, an upside down, inside out modern reimagining of the classic flood myth and a fictionalized origin story of Atmospheric Water Generation, an unexpected creative pivot hit, like a bolt of lightning; *TMY*'s sequel, *Bandwidth,* arrived almost fully formed, its title, its characters names, the book cover design, even its structural arc. And written just days before *Bandwidth* appeared in my mind's eye a poem, titled *Colorless,* surfaced like a battle cry. Poetry would be my creative side kick throughout the writing of *Bandwidth* and the following poem started the flood of words.

Constantly amazed by the unexplained ability of humans to channel creative impulse into form, both the poem and novel speak directly to our shared humanity, not only in the face of dehumanizing technologies, but more importantly the dehumanizing acts we perpetrate against one another.

I do hope you enjoy them both and feel compelled to help stop the unnecessary pain through random acts of kindness.

Tanya H. Van Cott

November 2025

TANYA H. VAN COTT

COLORLESS

by Tanya Van Cott

6/12/2021

Color Theory says, "…"
No, they were all liars!
A value system, shades of grey,
Or isn't it brown?

Pigment,
without it,
in art and skin,
are all things equal?

Judged by the value of a color?
A mere shield,
From the extraterrestrial ultraviolet?

Different cultures,
Different values,
Different proximities to the fiery ball we revolve around.
Yet, skin-deep,
Even at mid-scale,
There is judgement.
Too much pigment judged as an epic fail?

Colorless,
Is the Art…
…World,
I seek.

Can we train the eye to unsee?
And finally value each other?

TANYA H. VAN COTT

TANYA H. VAN COTT

BANDWIDTH

TANYA H. VAN COTT

www.ingramcontent.com/pod-product-compliance
Lightning Source LLC
Chambersburg PA
CBHW030427310726
48979CB00009B/1646/J
9781969866012